MANUFACTURING TECHNOLOGY RESEARCH

AF764716

THE MANUFACTURING EXTENSION PARTNERSHIP PROGRAM

OVERVIEW AND ASSESSMENT

MANUFACTURING TECHNOLOGY RESEARCH

Additional books in this series can be found on Nova's website under the Series tab.

Additional e-books in this series can be found on Nova's website under the eBooks tab.

MANUFACTURING TECHNOLOGY RESEARCH

THE MANUFACTURING EXTENSION PARTNERSHIP PROGRAM

OVERVIEW AND ASSESSMENT

RANDOLPH S. PERKINS
EDITOR

NOVINKA

New York

Library of Congress Cataloging-in-Publication Data

ISBN: 978-1-63485-193-0

Published by Nova Science Publishers, Inc. † New York

CONTENTS

PREFACE

The Hollings Manufacturing Extension Partnership (MEP) program is a national network of centers established by the Omnibus Trade and Competitiveness Act. The program's objectives are to enhance productivity and technological performance and to strengthen the global competitiveness of target manufacturing firms, namely small and medium-sized U.S.-based firms. Under this program, National Institute of Standards and Technology (NIST) partners with 60 nonfederal organizations called MEP centers. The centers, located in 50 states and Puerto Rico, help target firms develop new customers and expand capacity, among other things. NIST awards federal funding to centers under annually renewed cooperative agreements, subject to the centers providing matching funds and receiving a positive performance evaluation. This book describes, over the past 5 years, how NIST spent federal MEP program funds and examines the basis for NIST's cooperative agreement award spending.

In: The Manufacturing Extension Partnership … ISBN: 978-1-63485-193-0
Editor: Randolph S. Perkins

Chapter 1

THE MANUFACTURING EXTENSION PARTNERSHIP PROGRAM*

John F. Sargent Jr.

SUMMARY

The Hollings Manufacturing Extension Partnership (MEP) program is a national network of centers established by the Omnibus Trade and Competitiveness Act (P.L. 100-418). MEP centers provide custom services to small and medium-sized manufacturers (SMMs) to improve production processes, upgrade technological capabilities, and facilitate product innovation. Operating under the auspices of the National Institute of Standards and Technology (NIST), the MEP system includes centers in all 50 states and Puerto Rico.

The MEP program received $130.0 million in appropriations for FY2016, equal to its FY2015 level and $11 million less than the President's request. NIST provides funding to support center operations, with matching funds provided by nonfederal sources (e.g., state governments, fees for services). Initially established with a goal of transferring technology developed in federal laboratories to SMMs, MEP shifted its focus in the early 1990s to responding to needs identified by SMMs, including off-the-shelf technologies and business advice. As MEP evolved, its focus shifted to reducing manufacturing costs through

* This is an edited, reformatted and augmented version of a Congressional Research Service publication R44308, prepared for Members and Committees of Congress, dated January 12, 2016.

lean production, quality, and other programs targeting plant efficiencies and to increasing profitability through growth. Current MEP efforts focus on innovation strategies, commercialization, lean production, process improvements, workforce training, supply chain optimization, and exporting.

In 2014, MEP began a system-wide revamp intended to align center funding levels more closely with the national distribution of manufacturing activity; allow a federal cost-share of up to 50% for the first three years of each center's new cooperative agreement; and result in a single center in each state and Puerto Rico. Other objectives include aligning center activities to the NIST MEP strategic plan; aligning center activities with state and local strategies; providing opportunities for new partnering arrangements; and restructuring and reinvigorating the boards of local centers.

The MEP program has, at times, been included in discussions surrounding termination of federal programs that provide direct support for industry. Proponents assert that SMMs play a central role in the U.S. economy and that the MEP system provides information and assistance not otherwise available to SMMs. Some opponents have asserted that such services are available from other sources and that MEP inappropriately shifts a portion of the costs of these services to taxpayers.

Continued federal support for MEP centers remains a point of contention. As conceived, the centers were intended to become self-supporting after six years. The original legislation provided for a 50% federal cost-share for the first three years of operation, followed by declining levels of federal support for the final three years. Federal funding after a center's sixth year of operation was prohibited. In 1998, Congress eliminated the prohibition on federal funding after year six. Invoking the intent of the original legislation, the George W. Bush Administration proposed in its FY2009 budget to eliminate federal funding for MEP and to provide for "the orderly change of MEP centers to a self-supporting basis." Congress has continued to appropriate funding for MEP.

A related issue is the level of the federal cost-share for the centers. Currently, centers may receive a 50% federal cost-share in their first three years of operation, a 40% cost-share in year four, and a one-third cost-share in their fifth and subsequent years. Some MEP advocates would like the federal government to provide up to 50% of center costs, regardless of how many years a center has been in operation, to allow centers to reach SMMs they might not otherwise be able to serve. The ongoing system-wide competition of the centers will essentially reset the clock, allowing centers to receive a 50% cost-share for the first three years of their new cooperative agreements.

As Congress makes appropriation decisions, it may continue to discuss support for MEP in the context of the federal government's role in facilitating technological advancement and bolstering innovation and competitiveness.

Overview

The Hollings Manufacturing Extension Partnership (MEP), a program of the National Institute of Standards and Technology (NIST),[1] is a national network of centers that provide custom services to small and medium-sized manufacturers (SMMs)[2] to improve production processes, upgrade technological capabilities, and facilitate product innovation.

The MEP mission is "to enhance the productivity and technological performance of U.S. manufacturing." The MEP program executes this mission through "state and regional centers [that] facilitate and accelerate the transfer of manufacturing technology in partnership with industry, universities and educational institutions, state governments, and NIST and other federal research laboratories and agencies."[3] Funding for the MEP centers is provided on a cost-shared basis between the federal government and nonfederal sources, including state and local governments and fees charged to SMMs for center services.[4]

The MEP program received $130.0 million in appropriations for FY2016, equal to its FY2015 level and $11 million less than the President's FY2016 request. The MEP had a staff of 49 employees at NIST in FY2015,[5] and the centers have approximately 1,300 field staff with technical and business expertise. MEP is currently engaged in a system-wide four-round competition that will award one center to each state and Puerto Rico; currently some states have more than one MEP center. To date, two rounds of the competition have been completed, resulting in awards to centers in 20 states. Two additional rounds are planned for FY2016.

NIST served more than 30,000 SMMs in FY2014. In a survey of clients, NIST found that companies reported $2.5 billion in new sales, $4.2 billion in retained sales, $1.1 billion in cost savings, $2.7 billion in new client investment, the creation of 17,833 jobs, and the retention of 46,069 jobs in FY2014.[6] Fifty-eight percent of MEP clients receiving in-depth technical assistance reported increases in sales, reduction in costs, or new investments resulting from the services received.[7]

BACKGROUND

In the mid-1980s, congressional debates on trade focused attention on the critical role of technological advance in the competitiveness of individual firms and long-term national economic growth and productivity. Reflecting these ideas, the Omnibus Trade and Competitiveness Act (P.L. 100-418) established a public-private program, now known as the Hollings Manufacturing Extension Partnership, to assist U.S.-based SMMs in identifying and adopting new technologies. The focus on SMMs derived from their perceived contribution to job creation, innovation, and manufacturing. Research at that time indicated that SMMs produce 2.5 times more innovations per employee than large firms.[8] Program advocates noted the efforts of other nations to provide technical and business assistance to their manufacturing communities through the establishment of manufacturing extension centers (see text box, "MEP-Like Programs of Other Countries").

MEP-Like Programs of Other Countries

Several other countries also have national networks of centers that provide technical and business support to small and medium-sized manufacturers. For example:

- Japan's Kohsetsushi network received $2.140 billion
- in 2012 and has 182 centers and 6,000 technical staff.
- Germany's Fraunhofer Institutes received $595
- million in 2012 and has 60 centers and 15,000 technical staff.
- Canada's Industrial Research Assistance Program
- (IRAP) received $258 million in government funding and has more than 100 centers and more than 200 technical staff.

Like the MEP, the Fraunhofer Institutes and at least some of the Kohsetsushi centers charge clients fees for their services; IRAP does not charge clients.

Source: U.S. Government Accountability Office, Global Manufacturing: Foreign Government Programs Differ in Some Key Respects From Those in the United States, GAO-13- 265, July 2013.

In 2012, there were 253,000 SMMs in the United States. These firms accounted for nearly 99% of the nation's manufacturing enterprises and employed approximately 5.1 million people in 2012, approximately 45% of total U.S. manufacturing employment.[9]

The improved use of technology by SMMs is seen by policymakers and business analysts as important to the competitiveness of American manufacturing firms. How a product is designed and produced often determines costs, quality, and reliability. Lack of attention to process technologies and techniques may be the result of various factors, including company finances, insufficient information, equipment shortages, and undervaluation of the benefits of technology. A key purpose of the MEP program is to address these issues through outreach and the application of expertise, technologies, and knowledge.

NIST requires regular reporting by the centers, including the number and types of projects undertaken. Centers also are mandated to collect information from client companies that may provide indicators of longer-term results, including changes in sales, financial investments, inventory reduction, savings in labor and materials, and jobs created or saved. According to NIST, from MEP's inception through FY2014, the program has worked with nearly 80,000 manufacturers, leading to $88 billion in sales and $14 billion in cost savings, and has helped create more than 729,000 jobs.[10]

According to NIST, for every dollar of federal investment, the MEP generates nearly $21 in new client investment and $19 in new sales growth for SMMs. NIST also asserts that MEP creates or retains one manufacturing job for every $1,978 in federal investment.[11]

Evolution of the Program

The MEP program was originally established in 1988 as the "Regional Centers for the Transfer of Manufacturing Technology."[12] Over time, the program was referred to by a number of different names, including the Manufacturing Technology Centers program and the Manufacturing Extension Partnership program. The America COMPETES Reauthorization of 2010 codified the name of the program as the "Hollings Manufacturing Extension Partnership" and the centers as the "Hollings Manufacturing Extension Centers."[13]

From its inception through the mid-1990s, the MEP's principal emphasis was on

> establishing the national network—making sure there was a center within reach of all the nation's manufacturers and linking those centers to one another so they could learn from and teach each other about how best to work with manufacturers.[14]

The first three centers were established in 1989. Four more were added in 1991 and 1992. In 1994, the number of MEP centers expanded substantially when NIST took over support of extension centers originally funded by the Department of Defense's Technology Reinvestment Project. This brought the number of centers to 44. NIST awarded additional centers in 1995-1996, increasing the total to 70 centers.[15] Subsequent consolidation of centers in New York and Ohio brought the number of centers down to 60, including centers in each state and Puerto Rico.

While the focus on helping SMMs has remained constant, the methods and tools used by MEP have evolved since its creation. An intent of the legislation that created the manufacturing extension effort was to provide cutting-edge technology developed by NIST and other federal laboratories to SMMs. Royalties and licensing fees paid to the centers by the SMMs for the use of these technologies were expected to make the centers self-sufficient after the initial six years of operation. Advanced, federally funded technology, however, did not prove to be what most SMMs needed. Rather, their needs proved to be much more basic, including off-the-shelf technologies and business advice on topics such as management information technology, financial management systems, and business processes. A 1991 assessment of the program by the General Accounting Office (GAO, now the Government Accountability Office) concluded that

> While legislation establishing the Manufacturing Technology Centers Program emphasized the transfer of advanced technologies being developed at federal laboratories, the centers have found that their clients primarily need proven technologies. *Thus, a key mandate of this program is not realistically aligned with the basic needs of most small manufacturers* [emphasis added]... [A]ccording to officials from professional and trade associations representing small manufacturers and the results of key studies on U.S. manufacturing competitiveness, such advanced, laboratory-based technologies are not practical for most small manufacturers because these technologies generally are expensive, untested, and too complex.[16]

In recognition of this situation, the program was reoriented to offer more basic technologies that helped SMMs to improve their productivity and competitive position. By the mid-1990s, MEP was providing "a wide range of business services, including helping companies (1) solve individual manufacturing problems, (2) obtain training for their workers, (3) create marketing plans, and (4) upgrade their equipment and computers."[17] As articulated in the NIST Manufacturing Innovation blog,

> The initial services were focused on solving immediate and short-term problems—point solutions. The philosophy was an engineering one: 'You have a problem. We can fix it.'[18]

Over time, the MEP's focus moved from point solutions to more strategic, integrated services. In 2010, the "overarching strategy" for the MEP program was to reduce manufacturing costs through "lean, quality, and other programs targeting plant efficiencies" and to increase profitability "through business growth services resulting in new sales, new markets, and new products."[19]

Current MEP efforts focus on innovation strategies, commercialization, lean production, process improvements, workforce training, supply chain optimization, and exporting. One of the key areas of the MEP strategy is technology acceleration.[20] MEP defines technology acceleration as

> integrating technology into the products, processes, services and business models of manufacturers to solve manufacturing problems or pursue opportunities and facilitate competitiveness and enhance manufacturing growth. Technology Acceleration spans the innovation continuum and can include aspects of technology transfer, technology transition, technology diffusion, technology deployment and manufacturing implementation.[21]

Technology acceleration encompasses MEP efforts to assist SMMs in the improvement of existing products, the development of new products, and the development and improvement of manufacturing processes. MEP assists SMMs in this regard through a variety of approaches including technology scouting and transfer; supplier scouting; business-to-business network pilots; lean product development; technology-driven market intelligence; access to capital; cooperative research and development activities with NIST laboratories; and use of other federal programs such as the Small Business Innovation Research (SBIR) program,[22] the Advanced Manufacturing

Technology (AmTech) Consortia program, and the National Network for Manufacturing Innovation (NNMI).[23]

While continuing to offer its services to all SMMs, MEP is emphasizing targeted outreach toward growth-oriented SMMs and small entrepreneurial startups.[24]

MISSION

The statutory objective of the MEP centers is to enhance productivity and technological performance in U.S. manufacturing through the following:

- the transfer of manufacturing technology and techniques developed at [NIST] to Centers and, through them, to manufacturing companies throughout the United States;
- the participation of individuals from industry, universities, State governments, other Federal agencies, and, when appropriate, [NIST] in cooperative technology transfer activities;
- efforts to make new manufacturing technology and processes usable by U.S.- based small- and medium-sized companies;
- the active dissemination of scientific, engineering, technical, and management information about manufacturing to industrial firms, including small- and medium-sized manufacturing companies;
- the utilization, when appropriate, of the expertise and capability that exists in Federal laboratories other than [NIST]; and
- providing to community colleges information about the job skills needed in small- and medium-sized manufacturing businesses in the regions they serve.[25]

No direct financial support is available for companies through the centers. The program offers only technical and managerial assistance, and the cost of that assistance is generally reimbursable on a sliding scale.[26]

MEP ORGANIZATION AND STRUCTURE

The MEP program includes an MEP program office located at NIST (NIST MEP), an MEP Advisory Board, and the MEP centers.

NIST MEP

The NIST MEP program office is led by a director and has six components:

- Center Operations is responsible for providing financial and programmatic oversight for federal funding awarded to support the MEP mission; providing cooperative agreement and operational assistance and guidance to MEP centers; and supporting the MEP system of centers in partnership with NIST MEP's Regional Managers for Strategic Transitions and NIST Grants Management Division.
- Partnerships and Program Development is responsible for developing and maintaining partnerships and creating and launching programs to improve the services offered by MEP centers. Partnerships and Program Development helps identify and develop new opportunities with and for centers to help their clients, and helps identify, develop, and maintain partnerships of national significance.
- Manufacturing Policy and Research conducts performance evaluations for the MEP center system and facilitates reporting of MEP performance data.
- System Operations Office helps MEP centers identify opportunities for serving manufacturers with a focus on profitable growth. The team works with MEP center, state, and industry leaders to support the development of partnerships focused on local manufacturing ecosystems.
- Communications is responsible for messaging and outreach efforts focused on highlighting MEP accomplishments and positioning the program as a resource for manufacturers. Communications works with the local MEP centers on branding and marketing efforts, coordinates the efforts of the MEP Advisory Board, shares program information with the general public, and responds to inquiries from other stakeholders.
- Administration and Finance is responsible for providing internal and external customer service; overall management of administrative functions, budget/finance, human resources, center reviews, information technology (IT) support and security; and property management necessary for effective and efficient operations

throughout the MEP program. Administration and Finance provides advice and support to the MEP director and MEP staff to ensure proper understanding and oversight of administrative, IT, and financial issues.[27]

In FY2015, NIST MEP had 49 employees and received appropriations to support 80 FTE.[28] The NIST FY2016 budget justification requested funding for 80 FTE and authorization of 81 for MEP.[29]

MEP Advisory Board

Congress established an MEP Advisory Board to provide the NIST Director with advice on MEP programs, plans, and policies; assessments of the soundness of MEP plans and strategies; and assessments of current performance against MEP program plans.[30] By statute, the MEP Advisory Board is to consist of 10 members broadly representative of stakeholders appointed by the NIST Director. The board is to include at least two members employed by or on an advisory board for a center, and at least five members from U.S. small businesses in the manufacturing sector. Federal employees may not serve as advisory board members. Members serve staggered terms of three years. A member may serve two consecutive terms. One year from the end of the second term, a member may be re-appointed to the board.

The MEP Advisory Board is to act solely in an advisory capacity in accordance with the Federal Advisory Committee Act.[31] The board is required to meet at least twice a year and to report annually to Congress, through the Secretary of Commerce, on the status of the MEP program and programmatic planning. Copies of the MEP Advisory Board annual reports are available online at http://www.nist.gov/mep/about/advisory-board-reports.cfm.

MEP Centers

The MEP program is administered by NIST through partnerships with centers in all 50 states and Puerto Rico, including approximately 400 service locations[32] and nearly 1,300 field staff with technical and business expertise.[33] MEP seeks to have a center or other service location not more than two hours away from any potential client. A complete list of current MEP centers is provided in *Appendix A*.

Each center is operated by a state government, university, or other nonprofit organization. Center staff are employees of the center and its partners, not the federal government.

CENTER SELECTION

The following sections provide an overview of the criteria used by NIST MEP in awarding centers and the ongoing system-wide center competition.

Criteria

MEP centers are selected in response to open and competitive solicitations issued by NIST. Federal statute requires that center selections be based on merit using, at a minimum, the following criteria:

- the merits of the application, particularly those portions of the application regarding technology transfer, training and education, and adaptation of manufacturing technologies to the needs of particular industrial sectors;
- the quality of service to be provided;
- geographical diversity and extent of service area; and
- the percentage of funding and amount of in-kind commitment from other sources.[34]

System-Wide Center Competition

Following the first MEP center awards in 1989, the number of centers grew to 70, including at least one center in each state and Puerto Rico, and two or more centers in a few states. Later consolidation reduced the number to 60.

In 2013, Congress directed the Government Accountability Office to examine the basis for NIST's cooperative agreement award spending on the centers.[35] In March 2014, the GAO concluded that federal funding for MEP centers was imbalanced based on the number and industrial composition of manufacturers to be served and the cost of serving SMMs in different parts of the country.[36] GAO noted that NIST MEP funding through cooperative agreement awards to centers initially took into account factors such as the number of SMMs in the service area; the characteristics of the SMMs, including business size, industry types, product mix, and technology requirements; and the cost of providing services to those firms. However, GAO concluded that since these awards were made over 15 years, the allocation of funds in 2013 didn't appropriately account for these factors. Accordingly, GAO concluded that

> NIST's cooperative agreement award spending may not allow centers to provide the same level of services to target manufacturing firms, according to their needs.[37]

GAO recommended that

> [NIST's] spending on cooperative agreement awards be revised to account for variations across service areas in demand for program services and in MEP centers' costs of providing services.[38]

In its response letter, NIST agreed with GAO's conclusion and recommendation, and identified avenues it was exploring to take into account such variations. NIST estimated that if additional funding were provided to underfunded centers, the number of SMMs served in those service areas would increase by up to 20% over three years. GAO noted that such increases in funding for underfunded centers could result in decreases for other centers and noted the option of phasing in changes gradually to minimize disruption to centers and the SMMs they serve. NIST expressed concerns that some centers may not be able to access the additional federal funds due to difficulties in meeting the 2-to-1 nonfederal match required under current funding requirements.[39]

The GAO report, together with recommendations of the MEP Advisory Board and other factors, contributed to NIST's decision to develop a strategy for executing a system-wide center competition via four multi-state competitions beginning in 2014 and concluding in 2017. According to NIST, the system-wide competition will result in center funding levels more closely reflecting the national distribution of manufacturing activity; allow a federal cost-share of up to 50% for the first three years of each center's new cooperative agreement; and result in a single center in each state and Puerto Rico. Other objectives include aligning center activities to the NIST MEP strategic plan; aligning center activities with state and local strategies;

providing opportunities for new partnering arrangements; and restructuring and reinvigorating local center boards.[40]

For most states, centers will be awarded in one of the four rounds of competitions. A few states with centers that were competed in recent years will not take part in the system-wide center competition; however, the terms of their cooperative agreements will be updated to align with those of the other centers.

The first round of center competitions included the 10 states that were deemed most underfunded in terms of dollars per manufacturing establishment.[41] Awardees from the first competition were announced in February 2015. The second round was announced in September 2015. In total, NIST has awarded 20 centers under these competitions, with one center award pending.[42]

According to NIST,

> Proposals were reviewed by government and independent experts and evaluated against a number of criteria, including demonstration of a thorough understanding of market needs and how proposed service offerings would meet those needs. The reviewers also looked at the proposed business models, performance measurements and metrics, partnership potential, staff qualifications and program management, as well as financial and nonfederal cost-share plans.[43]

Through the first two rounds of the system-wide competition few states had more than one applicant. In round one, only 2 of the 10 states had more than one applicant. In round two, only 3 of the 11 states had more than one applicant; no applications were received for the Ohio competition.[44] Each of the 19 center awardees selected to date in the first two rounds have been incumbent MEP centers. Despite few states having more than one applicant and all selections being incumbent centers, NIST notes:

> Given the complex nature of this program, many competing applicants seek to partner with other organizations to strengthen their proposals. The challenge of a full and open competitive process required the centers to substantially enhance programs, partnerships, and funding sources, fulfilling the goal of getting the most out of MEP's investment by providing it to those organizations best qualified to deliver services to small- and medium-sized manufacturers.[45]

In addition, NIST has noted that despite the limited competition for the centers, the number of partners involved with the centers that have been awarded increased by 32%. NIST asserts that it is applying lessons learned from the first two rounds to improve the competition process in future rounds.[46] In particular, NIST MEP is continuing outreach efforts—including working with stakeholders through their newsletters and websites—to increase awareness of the opportunity.[47]

NIST expects to begin the third round of competition in January 2016 and the fourth round in July 2016.[48] NIST anticipates completing the process by 2017, subject to availability of funds.[49] NIST seeks to accomplish this transition without disrupting ongoing local service to SMMs or degrading the performance of the national MEP system.

Review Prior to Continued Center Funding

Center awards are made as cooperative agreements with an initial performance period of five years. NIST may extend an award for an additional five years following an overall assessment of the center, including "programmatic, policy, financial, administrative, and responsibility assessments."[50] According to NIST, when an application for a multi-year award is approved, funding is usually provided for only the first year of the project; for subsequent years, recipients are required to submit detailed budgets and budget narratives prior to the award of any continued funding. The amount of funds awarded after the first year is provided on a noncompetitive basis and may be adjusted upward or downward. Center funding after the first year is contingent upon satisfactory performance, continued relevance to the mission and priorities of the program, and the availability of funds. Continuation of an award to extend the period of performance or to increase or decrease funding is at the sole discretion of NIST.[51]

Center Cost-Share and Term of Eligibility

The following sections provide an overview of center cost share requirements and term of eligibility, as well as allocation of MEP center funding, including the current status, a discussion of the historical background and ongoing issues, and the effect of the system-wide competition on center cost share.

Current Status

Funding for the MEP centers is provided on a cost-share basis by the federal government and nonfederal sources. NIST may provide no more than one-half of center costs during the first three years of an award, no more than two-fifths in the fourth year, and no more than one-third in year five and beyond. (See *Table 1.*) While centers may take on a wide range of activities, for purposes of these calculations, center costs include those incurred in connection with activities undertaken to improve the management, productivity, and technological performance of SMMs.[52] There is currently no limit to the number of years a center may receive federal funding.

Table 1. MEP Center Federal and Nonfederal Cost-Share

Award Year	Maximum NIST Cost-Share	Minimum NonfederalCost-Share
Years 1-3	1/2	1/2
Year 4	2/5	3/5
Year 5 and beyond	1/3	2/3

Source: NIST, "Award Competitions for Hollings Manufacturing Extension Partnership (MEP)," 79 Federal Register 44746-44752 , August 1, 2014, https://federalregistergov/a/2014-18264.

As discussed above, the ongoing system-wide competition is intended to better align center funding levels with the number of SMMs and the cost of providing services to these firms in each center's service area. In this regard, NIST MEP has set federal funding levels for each state center. These amounts are the maximum available for the federal cost-share, and a center must meet the required nonfederal cost-share to be eligible to receive full funding. The federal share of funding for MEP centers made in the two rounds awarded in 2015 ranged from $500,000 to $6.7 million per year. (*Appendix B* provides annual funding awarded centers in each state from the first two rounds of competition.)

Historical Background and Ongoing Issues

The following sections provide information on centers' term of eligibility for funding, cost-sharing requirements, and the effect of the ongoing system-wide competition on cost-sharing.

Term of Eligibility for Funding

The legislation that established the MEP program initially prohibited centers from receiving federal financing beyond their sixth year of operation.[53] However, federal support beyond the sixth year later became considered necessary in lieu of increasing service charges paid by SMMs. While analysts considered service charges to the SMMs to be important to the effectiveness of the MEP program,[54] some also expressed concerns that an increase in charges commensurate with making the centers self-supporting might make the services too expensive for many SMMs. This perspective was articulated in a 1998 NIST-sponsored study, Analysis indicates that to offset lost public revenue centers would need to take on much larger projects at much higher billing rates and focus on repeat business. As a result,

> many small manufacturers would not be able to afford these services. Given this conclusion, the best way to ensure high-caliber nationwide assistance to smaller manufacturers is to commit to a stable amount of renewable federal funding for those centers which receive successful evaluations.[55]

The prohibition on funding after the sixth year was temporarily suspended by provisions in the FY1997 and FY1998 appropriations acts,[56] then eliminated by the Technology Administration Act of 1998 (Section 2, P.L. 105-309). Under the provisions of the act, centers were eligible to receive federal funding of up to one-third of center costs after their sixth year of operation, subject to positive, independent evaluations to be conducted at least every two years.

Cost-Sharing

The financial support system created for MEP by Congress in the original legislation was based on matching financing between the federal government and state, local, and/or private nonprofit entities. The Senate Committee on Commerce, Science, and Transportation report to accompany the Technology Competitiveness Act of 1987 (S. 907, 100th Congress) directed that "the percentage of funding offered by particular applicants be considered in deciding which applications be selected."[57] Cost-sharing strengthens the ties between the organizations involved in the cooperative arrangement and as such, the committee stated that "special attention will be given to innovative ways in which Federal laboratories, State agencies, and business and professional groups can work together."[58] The matching provisions were seen as a means to ensure that the centers reflect the actual needs of the manufacturing companies in the area they serve.

Following the economic downturn of 2007-2009, there were calls for Congress to raise the federal cost-share to 50%. At that time, some commentators argued that during the difficult economic situation, state and local financial support for the program may be curtailed. At the same time, client fees for service decreased 13.4% between FY2008 and FY2009, the first significant decline since FY1996.[59] Advocates of increasing the federal share noted that such action would not release state and local partners of their responsibility to support the centers, but would permit continued outreach to small manufacturers without pricing the services out of reach. Opponents of this approach argued that the one-third federal contribution was sufficient and that the successful operation of the program was dependent on the financial participation of state and local government as well as the companies utilizing the centers.

The America COMPETES Reauthorization Act of 2010 (P.L. 111-358) mandated that the GAO explore and report on the cost-share provisions of the MEP program. In response, GAO issued a report on April 4, 2011, that noted:

> We were unable to provide recommendations on how best to structure the cost-share requirement to provide for the long-term sustainability of the program because we could not identify criteria or a basis for determining the optimal cost-share structure for this program. Instead, we have identified a number of factors that could be taken into account in considering modifications to the current cost-share structure. Among other things, past GAO work has found that cost-share structures should promote equity by assigning costs to those who both use and benefit from the services. As it applies to the MEP program, manufacturers, state and local governments, and the nation may all benefit from the program to varying degrees, requiring an evaluation of the relative benefits and aligning cost-shares to reflect who receives the benefits.[60]

In this regard, GAO noted that NIST's study of the cost-share provision of the MEP program

> recommended that the cost-share requirements should be consistent with those of other economic development programs—which it noted, in Commerce, had 1:1 or lower costsharing—and should provide flexibility to alter the cost-share requirement in response to economic conditions.[61]

However, GAO also noted that the Congressional Budget Office (CBO) had identified the MEP program for potential elimination from discretionary spending, stating that the program's enhancement of U.S. productivity is questionable. According to CBO, the legislative agency "regularly issues a compendium of budget options to help inform federal lawmakers about the implications of possible policy choices."[62] Elimination of MEP was one more than 100 options CBO proposed in 2011 for changes to federal spending and revenues.

In 2014, two bills were introduced with provisions that would have allowed federal support for MEP centers of up to 50% of annual costs incurred, without regard to how long the cooperative agreement has been in effect.[63] The NIST Reauthorization Act of 2014 (H.R. 5035, 113th Congress) passed the House but did not advance in the Senate. The America COMPETES Reauthorization Act of 2014 (S. 2757, 113th Congress) was introduced in the Senate but did not advance out of Committee.

Also in 2014, the MEP Advisory Board recommended that MEP readjust the cost-share structure in order to optimize the federal investment and provide for the long-term sustainability of the program. Specifically, the board recommended requiring to a 1:1 match (50% federal cost share) and allowing the nonfederal cost-share to include in-kind contributions of up to one-half of the center's portion of the cost-share.[64]

In 2015, the Senate Committee on Appropriations expressed concerns about the federal cost-share structure (as it existed prior to the recent system-wide competition) and directed NIST to provide a report to the committee and to the Senate Committee on Commerce, Science, and Transportation "detailing quantifiable metrics on total MEP center funding, including a breakdown of the type of contribution source across centers that have transitioned from the 50 percent Federal, 50 percent non-Federal cost-share to a lower cost-share held by the Federal Government."[65]

Effect of System-Wide Competition on Cost-Share

Since most MEP centers were in or past the fifth year of their cooperative agreements prior to the start of the current system-wide competition,[66] the federal cost-share was generally limited to one-third of center costs. The ongoing system-wide center competition essentially resets the clock on these centers, raising the maximum federal cost-share to one-half for the first three years that the new center agreements are in place, even if a center was funded under a previous agreement. NIST argues that the increased share of federal funding will allow centers to serve SMMs that they otherwise couldn't afford to serve.

OTHER MEP-RELATED ACTIVITIES

The MEP program has provided additional funding opportunities for a number of activities that support the program's overarching mission. Some of these activities were supported solely by NIST, while others were supported by multiple federal agencies. Recent activities of this type include business-to-business networks, Make it in America Challenge, Advanced Manufacturing Jobs and Innovation Accelerator Challenge, and Manufacturing Technology Acceleration Centers.

Business-to-Business Networks

In December 2014, NIST MEP awarded $2.5 million to 10 MEP centers for the establishment of pilot projects to develop, deploy, and maintain business-to-business (B2B) networks.[67] These networks are intended to help match buyers and sellers of technologies or products and services in support of SMMs. The two-year projects are designed to be scalable and interoperable to help determine whether they could be expanded into a national network or a series of regional ones.[68]

Make it in America Challenge

In December 2013, NIST MEP awarded grants to 10 winners in nine states as part of the multi-agency Make it in America (MiiA) Challenge, an

Obama administration initiative to accelerate job creation and encourage business investment in the United States. Nine awards were to MEP centers. Two were to affiliates of the Ohio MEP center. Each received $125,000 per year for three years.[69]

According to NIST, the program supports the efforts of U.S. companies to keep, expand, or re-shore manufacturing operations and jobs in the United States, and to encourage foreign companies to build facilities in the United States and make products domestically. The MEP's MiiA Challenge grants are intended to support greater connectivity in regional supply chains and to assist SMMs.

Advanced Manufacturing Jobs and Innovation Accelerator Challenge

NIST MEP centers are participating in the Advanced Manufacturing Jobs and Innovation Accelerator Challenge (AMJIAC), a multi-agency effort seeking to strengthen U.S. manufacturing.[70] A 2012 solicitation led to 10 three-year awards totaling $20 million.

According to NIST:

> These grants support the creation and strengthening of regional partnerships capable of accelerating innovation and growing a region's capacity for advanced manufacturing. This funding has been used for activities such as worker training programs or connecting manufacturers to resources like national labs or universities. Ultimately, these grants present regions with an opportunity not only to expand their current activities, but also to fundamentally transform the way that the region supports its manufacturers.[71]

The role of the MEP center participation varies in the awards. In some cases, an MEP center has the primary management role. In other cases, an MEP center is engaged in a partnership with another organization to lead different project elements. In still other cases, an MEP center is part of a broad-based partnership with different organizations leading one or two project elements.

Manufacturing Technology Acceleration Centers

In July 2013, NIST announced a new pilot program under MEP, the Manufacturing Technology Acceleration Centers (M-TACs). M-TACs are designed

> to explore different approaches to providing manufacturers with the technology transition and commercialization assistance they need to compete successfully and grow their market share within manufacturing supply chains.[72]

As of February 2015, there were five M-TAC pilots, each led by one of the MEP centers. The MTAC pilots will operate through the end of calendar year 2015.

Additional Grants

In October 2010, NIST announced $9.1 million in cooperative agreements for 22 projects "designed to enhance the productivity, technological performance and global competitiveness of U.S. manufacturers."[73] The funding was provided by MEP on a competitive basis to nonprofit organizations to work with the MEP centers and address one or more of these areas identified by NIST as critical to U.S. manufacturing:

- responding to evolving supply chains;
- accelerating the adoption of new technology to build business growth;
- implementing environmentally sustainable processes;
- establishing and enabling strong workforces for the future; and
- encouraging cultures of continuous improvement.[74]

According to NIST, "The funding will help encourage the creation and adoption of improved technologies and provide resources to develop new products that respond to changing market needs." In this regard, the awards differed from other MEP center activities which do not support research activities.

MEP Strategic Plan

In 2014, NIST MEP began the process of developing a new strategic plan and produced a draft in November 2014.[75] Among other things, the plan identified the strategic goals and objectives. The four goals of the plan are: enhance the economic competitiveness of U.S. manufacturers; serve as a voice to and voice for manufacturers; support national, state, and regional, manufacturing ecosystems and partnerships; and develop MEP's capabilities as a learning organization and high performance system. More information can be found at http://www.nist.gov/mep/about/strategicplan.cfm.

Annual Report to Congress

NIST is required to annually produce and submit to Congress a three-year programmatic planning document, concurrent with the President's annual budget request. This report is to include an assessment of the NIST Director's governance of the MEP program. The latest version of the plan, *NIST Three-Year Programmatic Plan: 2016-2018,* can be accessed at http://www.nist.gov/director/prog-ofc/upload/3_year_plan_final_16_pages.pdf.

External Reviews and Recommendations

A number of organizations have reviewed and commented on the program's management and effectiveness, and some have offered recommendations for improving the program. The following sections discuss some of the findings and recommendations of these organizations.[76]

MEP Advisory Board

In its FY2014 annual report, the MEP Advisory Board reviewed NIST's efforts to respond to the board's earlier recommendations. The board recommended the establishment of a personnel exchange plan involving staff from the NIST MEP and the MEP centers. Such an exchange program, the board asserted, could help educate newer MEP staff members and provide a better understanding of daily center activities. In addition, the exchange plan could be used to train future center leaders, help in the development of national working groups, and provide an in-depth understanding of NIST processes and organization. A pilot program involving one NIST MEP staff member was undertaken in 2014.[77]

The board also recommended the establishment of a Center Advisory Group to provide advice to NIST MEP on matters such as center reporting burdens; center flexibility; program impacts, outputs, and outcomes; and program integrity. NIST MEP subsequently established a Center Advisory Group composed of two center directors from each of the six MEP regions. The group has focused on short-term improvements, new approaches to reporting and evaluations, expanding the definition of manufacturing, changes to MEP's evaluation system, project coding, reduction in reporting burdens, data sharing between centers within MEP's information system, and measurement of innovation projects.[78]

A third board recommendation was to codify, align, and integrate program evaluation with contract management. NIST MEP has reviewed existing contracts and developed processes for approving them. In addition, NIST MEP integrated a performance evaluation mechanism into the system.

Government Accountability Office

The Government Accountability Office has reviewed aspects of the MEP program on several occasions since the early 1990s.

In a March 2014 report, the GAO reported on its investigation into the extent to which the MEP program achieves administrative efficiencies. GAO found that 81.4% of MEP funding supported center awards with the balance devoted to contracts, staff, agency-wide overhead charges, and other items, some of which NIST considered direct support and some of which NIST considered administrative spending. In total, NIST estimated that more than 88.5% of federal MEP program spending in FY2013 was for direct support, and the remainder supported MEP administration.[79]

In 2010 Congress directed the GAO to report on the cost-share structure of the MEP program and provide recommendations for how best to structure the cost-share requirement to provide for the long-term sustainability of the program.[80] GAO concluded that it was unable to provide such recommendations as it could not identify criteria or a basis for determining the optimal cost-share structure for this program.[81] However, GAO cited a number of factors that could be taken into account in modifying the existing cost-share structure including promoting equity by assigning costs to those who both use and benefit from the services. In this regard, GAO identified potential beneficiaries as manufacturers, state and local governments, and the nation and recommended an evaluation of the relative benefits and aligning cost-shares to reflect who receives the benefits.[82] (See "Cost-Sharing" for a further discussion of GAO's findings.)

In an August 1995 briefing paper, the GAO explored how small and medium-sized firms were served by various manufacturing extension efforts, including the MEP program.[83] GAO received 551 responses to 766 questionnaires distributed. Approximately 73% of responding firms stated that their relationships with an extension activity had a positive effect on the company's business performance. Fifteen percent indicated that there was no effect at all. Among the impacts identified were improved use of technology (63%), better product quality (61%), and expanded productivity (56%). According to GAO, this suggested that manufacturing extension activities "had some success in achieving their primary goal of helping manufacturers improve their operations through the use of appropriate technologies and through increases in product quality and worker productivity." The study also found that companies which used internal funding to implement recommendations offered by extension programs were the most likely to find an overall positive impact. "Significantly, approximately 97 percent of [these respondents] ... said that they believed that this investment had been worthwhile." Those who utilized these organizations noted that practical experience in the field contributed to the success of staff activities, as did the affordability of the assistance. Companies that did not utilize the resources provided by the MEP tended to be those that were unaware of the program and the opportunities associated with it.

Further refining this information in a March 1996 report, GAO also noted that company size and age were significant factors in business perceptions of the extension program. Smaller (under $1 million gross sales) and newer (established after 1985) firms "were most likely to report that their overall business performance was boosted by MEP assistance."[84] While there were no real differences in perception between extension services offered by NIST and those funded by other institutions, there was a difference in assessments of effectiveness based on whether or not payment was required. According to GAO, those firms that paid fees "were half as likely as those that paid no fees to credit the assistance for having an extremely positive impact, as opposed to a generally positive impact, on their business performance."

Congressional Budget Office

As discussed earlier, the CBO regularly issues a compendium of budget options to help inform federal lawmakers about the implications of possible policy choices. In 2009 and 2011, one of the options CBO proposed was elimination of the MEP program.

In its 2009 narrative, CBO asserted that proponents of elimination question the appropriateness and necessity of the type of technical assistance offered by MEP, stating that "many university professors of business, science, and engineering consult with private industry, and other ties between universities and business promote knowledge transfer," that many centers in the MEP system existed before the establishment of the MEP program, and that surveys indicated that about half of MEP's clients reported that the same services were available to them through other channels but at a higher price. Supporters of the MEP program, according to CBO, point to the importance of SMMs to the economy in terms of output and employment, and in providing supplies and intermediate goods for large companies. Proponents also argue that many SMMs "face barriers that can prevent them from obtaining the sort of information" that MEP provides.[85]

CBO also asserted that

> The program's enhancement of U.S. productivity also is questionable. It can be argued that federal spending for [MEP] allows some inefficient companies to remain in business, tying up capital, labor, and other resources that could be used more productively elsewhere.[86]

National Academy of Public Administration

The National Academy of Public Administration also studied the MEP program and in a 2004 report stated that while "on balance ... the MEP Program performs capably and effectively and that the core premise ... remains viable as it is fulfilling its mission by leveraging both public and private resources to assist the nation's small manufacturers," there should be consideration of a "fundamental change in the mix of the types of services it provides as well as the structures for delivering them."[87] As such, a Next Generation Strategic Plan was developed by the MEP in 2006 to concentrate on not just the shop floor but on "the entire enterprise and its position in the marketplace." In addition to individual manufacturing firms, NIST concluded that MEP "must focus on industry/supply chain requirements as well as overall economic development trends."[88] Current MEP efforts include a focus on helping companies to participate in supply chains (e.g., by helping them become compliant with quality standards) and on supply chain optimization.

APPROPRIATIONS AND RELATED ISSUES

The following sections provide information on the status of FY2016 appropriations for MEP and a longer term perspective on MEP budget requests and appropriations from FY2003-FY2016.

FY2016 Appropriations Status

Division B of the Consolidated Appropriations Act, 2016 (P.L. 114-113) provides $130.0 million for MEP for FY2016, the same as for FY2015. NIST requested $141.0 million for FY2016, $11 million above the FY2015 enacted level. NIST asserted that the additional funding "is necessary in order to optimize the impact of the federal investment on U.S. manufacturing" under the system-wide competition.[89] In June 2015, the House voted to appropriate $130.0 million for MEP in FY2016.[90] The Senate Committee on Appropriations subsequently recommended $130.0 million for MEP in FY2016 and stated its support for "MEP's focus on strengthening the existing network of MEP centers and providing additional support to centers based on the documented performance of the center's activities and the manufacturing capacity of the area served by the center."[91]

Appropriations and Requests FY2003-FY2016

The MEP program has at times enjoyed presidential and congressional support; at other times, it has been targeted for reductions or elimination. These changes are visible in the history of presidential budget requests and congressional actions on MEP appropriations. *Figure 1* illustrates funding levels for the NIST MEP program, both requested and enacted appropriations, for FY2003-FY2015; *Table 3* provides the requested and enacted appropriations amounts.

Table 2. Manufacturing Extension Partnership Program Appropriations, FY2016

(budget authority, in millions of dollars)

	FY2015 Enacted	FY2016 Request	FY2016 House	FY2016 S. Cmte.	FY2016 Enacted
Manufacturing Extension Partnership program	$130.0	$141.0	$130.0	$130.0	$130.0

Source: FY2016 Congressional Budget Justification for the National Institute of Standards and Technology/National Technical Information Service, http://www.osec.doc.gov/bmi/budget/FY16CJ/NISTNTIS_FY_2016_CJ_Final_508_Compliant.pdf; H.Rept. 114-130; H.R. 2578; S.Rept. 114-66; P.L. 114-113.

While President George W. Bush's annual budget requests generally called for substantial reductions in support for MEP, Congress appropriated generally steady funding except for FY2004 and FY2008. In FY2004, MEP funding was cut to $38.6 million, down 62.6% from its FY2003 level of $105.9 million. However, Congress restored MEP funding in FY2005, appropriating somewhat more than it had in FY2003.

In FY2008, MEP funding was cut to $89.6 million, down 14.4% from its FY2007 level of $104.7 million. For FY2009, President Bush's final budget proposed to end federal funding for MEP, requesting $4 million to allow for "the orderly change of MEP centers to a self-supporting basis."[92] Congress opted instead to provide $110.0 million for MEP, an increase of 22.8% above the FY2008 enacted level.

Under President Obama, MEP budget requests have equaled or exceeded actual appropriations. In FY2010, President Obama requested and received $124.7 million for MEP. Since then, the president's annual budget requests have proposed higher funding for MEP than has been enacted.

Between FY2005 and FY2015, MEP enacted appropriations have generally kept pace with inflation, growing at a compound annual growth rate (CAGR) of approximately 1.9% per year.[93]

Use of MEP Appropriations for Center Awards

In response to direction from Congress,[94] GAO investigated the extent to which the MEP program achieves administrative efficiencies. In its March 2014 report, GAO found that of the $608 million spent on the MEP program from FY2009 to FY2013, about $495 million (81.4%) went to center awards.

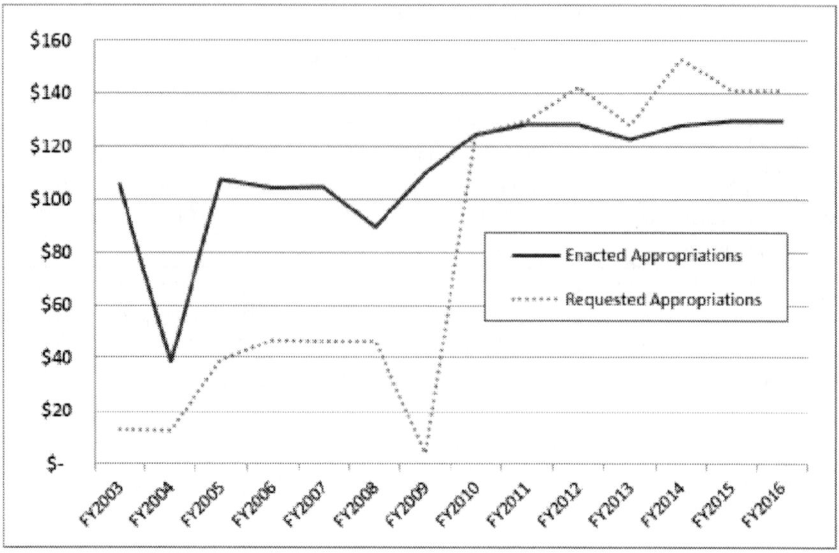

Source: Department of Commerce and NIST budget documents, FY2003-FY2016; P.L. 113-235; P.L. 114-113.

Figure 1. Manufacturing Extension Partnership Program Funding. Requested Appropriations, FY2003-2016; Enacted Appropriations, FY2003-FY2015 (in millions of current dollars).

Table 3. Requested and Enacted Appropriations for the MEP Program
(FY2003-FY2015, in millions of current dollars)

Fiscal Year	Request	Enacted
2003[a]	$ 12.9	$ 105.9
2004[b]	12.6	38.6
2005[c]	39.2	107.5
2006[d]	46.8	104.6
2007	46.3	104.7
2008	46.3	89.6
2009	4.0	110.0
2010	124.7	124.7
2011[e]	129.7	128.4
2012	142.6	128.4
2013[f]	128.0	123.0
2014	153.1	128.0
2015	141.0	130.0

Fiscal Year	Request	Enacted
2016	141.0	130.0

Source: Department of Commerce and NIST budget documents, FY2003-FY2016; P.L. 113-235; P.L. 114-113. Notes:

[a] Enacted levels reflect an across-the-board rescission enacted in P.L. 108-7.

[b] Enacted levels reflect across-the-board rescissions enacted in the FY2004 Consolidated Appropriations Act, P.L. 108-199, and NIST's share of the Department of Commerce's unobligated balances rescission.

[c] Enacted levels reflect across-the-board rescissions enacted in P.L. 108-447, FY2005 Consolidated Appropriations Act ($9.5 million). Does not reflect unobligated balances rescission of $3.9 million.

[d] Enacted levels reflect across-the-board rescissions enacted in P.L. 109-108, FY2006 Science, State, Justice, and Commerce Appropriations Act and in P.L. 109-148, FY2006 Defense Appropriations Act.

[e] Enacted levels include 0.2% across-the-board rescission.

[f] Enacted levels reflect the 1.877% rescission, 0.2% rescission, and the 5% sequester applied to 2013 annualized CR level.

The balance was spent on contracts, staff, agency-wide overhead charges, and other items, some of which NIST considered direct support and some of which NIST considered administrative spending. According to GAO, NIST estimated that more than 88.5% of federal MEP program spending in FY2013 was for direct support, and the remainder (11.5%) was for administration.[95]

Appropriate Role of the Federal Government

Continuing financial support for the MEP program is part of a larger ongoing debate among federal policymakers about the appropriate role of the federal government in providing assistance to U.S. industry. The MEP program has, at times, been included in discussions surrounding termination of federal programs that provide direct support for industry. Proponents assert that SMMs play a central role in the U.S. economy and that the MEP system provides information and assistance not otherwise available to SMMs. Some opponents have asserted that such services are available from other sources and that MEP inappropriately shifts a portion of the costs of these services to taxpayers. Proponents of the program stress that no direct funding is available to companies.

In addition, some have questioned whether federal support for the MEP centers should continue to be provided indefinitely. As originally expressed in statute, MEP centers were to receive no federal funding after their fifth year of operation, instead deriving necessary revenues from state and local governments as well as from the companies utilizing the center's services. In 1998, Congress lifted the prohibition on funding after the fifth year and allowed NIST MEP to provide up to one-third of center costs after their sixth year of operation indefinitely. More recently, some have argued for increasing the federal cost-share to one-half of center costs. The system-wide competition currently underway effectively accomplishes this, providing up to one-half of a center's costs for the first three years of the new awards. The debate over whether the federal government should continue to provide financial support to the centers indefinitely and, if so, at what level, may be revisited as the centers' federal funding decreases once again to two-fifths of the center costs in the fourth year of the new awards, and one-third in the fifth year.

These and other issues may be debated as Congress continues to make appropriation decisions relating to manufacturing extension as it pertains to the role of the federal government in facilitating research and technological advancement.

APPENDIX A. HOLLINGS MANUFACTURING EXTENSION PARTNERSHIP CENTERS

Table A-1. Hollings Manufacturing Extension Partnership Centers

State	Center Name, Address, and Website
Alabama	Alabama Technology Network 135 South Union Street, Suite 441, Montgomery, AL 36104 http://www.atn.org/
Alaska	Southwest Alaska Municipal Conference 3300 Arctic Boulevard,#203, Anchorage, AK 99503 http://www.swamc.org
Arizona	Arizona Commerce Authority 333 N. Central Avenue, Suite 1900, Phoenix, AZ85004 http://www.azcommerce.com\revaz
Arkansas	Arkansas Manufacturing Solutions 900 West Capitol Avenue, Suite 320, Little Rock, AR 72201 http://www.mfgsolutions.org
California	California Manufacturing Technology Consulting 690 Knox Street, Suite 200, Torrance, CA 90502. http://www.cmtc.com/

State	Center Name, Address, and Website
	Corporation for Manufacturing Excellence (Manex) 2010 Crow Canyon Place, Suite 320, San Ramon, CA 94583 http://www.manexconsulting.com
Colorado	Manufacturer'sEdge Manufacturer's Edge C/O REO, 5505 Airport Boulevard,Boulder, CO 80301 http://www.manufacturersedge.com
Connecticut	Connecticut State Technology Extension Program 1090 Elm Street, Suite 202, Rocky Hill, 06067 http://www.connstep.org/
Delaware	Delaware Technical and Community College 400 Stanton-Christiana Road, Suite A-158, Newark, DE 19713 http://www.demep.org/
Florida	FloridaMakes 800 N. Magnolia Avenue, Suite 1850, Orlando, 32803 http://www.floridamakes.com
Georgia	Georgia Manufacturing Extension Partnership Georgia Tech, 75 Fifth Street, NW Suite 300, Atlanta, GA30308 http://gamep.org/
Hawaii	INNOVATE Hawaii 2800 Woodlawn Drive, Suite100, Honolulu, HI 96822 http://www.innovatehawaii.org
Idaho:	Idaho TechHelp Boise State University, 1910 University Drive, Boise, 83725 http://www.techhelp.org
Illinois	Illinois Manufacturing Excellence Center –Chicagoland 1651 Wilkening Road, Schaumburg, IL 60173 http://www.imec.org
	Illinois Manufacturing Excellence Center –Downstate 428 Jobst Hall, 1501 W. Bradley Avenue,Bradley University, Peoria, 61625 http://www.imec.org
Indiana	Indiana MEP Purdue Technical Assistance Program 8628 E. 116thStreet, Suite 200, Fishers, IN 46038 http://www.mep.purdue.edu
Iowa	Iowa Center for Industrial Research and Service Iowa State University, Extension 4-H Building, Ames, IA 50011. http://www.ciras.iastate.edu
Kansas	Mid-America Manufacturing Technology Center 10550 Barkley Street,Suite 116, Overland Park, KS 66212 http://www.mamtc.com

Table A-1. (Continued)

State	Center Name, Address, and Website
Kentucky	Advantage Kentucky Alliance 2413 Nashville Road, B8, Suite 310, WKU Center for Research and Development, BowlingGreen, KY 42101. http://www.advantageky.org
Louisiana	Manufacturing Extension Partnership of Louisiana P.O. Box 53445, Lafayette, LA 70505 http://www.mepol.org
Maine	Maine Manufacturing Extension Partnership 87 Winthrop Street, Augusta, ME 04330 http://www.mainemep.org/
Maryland	Maryland MEP 8894 Stanford Boulevard, Suite 304, Columbia, MD 21045 http://www.mdmep.org
Massachusetts	Massachusetts Manufacturing Extension Partnership 100 Grove Street, Suite 108, Worcester, MA 01605 http://www.massmep.org/
Michigan	Michigan Manufacturing Technology Center 47911 Halyard, Plymouth, MI 48170 http://www.mmtc.org/
Minnesota	Enterprise Minnesota 310 4thAvenueS.,Suite 7050, Minneapolis, MN 55415 http://www.enterpriseminnesota.org
Mississippi	InnovateMEP Mississippi 134 Marketridge Drive, Ridgeland, 134 Marketridge Drive, Ridgeland, MS 39157 http://www.innovatemep.ms
Missouri	Missouri Enterprise 900 Innovation Drive, Suite 300, Rolla, MO 65401 http://www.missourienterprise.org
Montana	Montana Manufacturing Extension Center PO Box 174255, Montana State University, 2310 University Way Building2, Suite1, Bozeman,MT 59717. http://www.mtmanufacturingcenter.com
Nebraska	Nebraska Manufacturing Extension Partnership 210 L.W. Chase Hall, Lincoln, NE 68583 http://nemep.unl.edu
Nevada	Nevada Industry Excellence UNR Mail Stop 406, Reno, NV 89557 http://www.nevadaie.com
New Hampshire	New Hampshire Manufacturing Extension Partnership 172 Pembroke Road, Concord, NH 03301

State	Center Name, Address, and Website
	http://www.nhmep.org/
New Jersey	New Jersey Manufacturing Extension Program 2 Ridgedale Avenue, Suite 305, Cedar Knolls, NJ 07927 http://www.njmep.org
New Mexico	New Mexico Manufacturing Extension Partnership 4501 Indian School Road, NE, Suite 202, Albuquerque, NM 87110 http://www.newmexicomep.org
New York	New York Manufacturing ExtensionPartnership 625 Broadway, ESD, Division of Science, Technology & Innovation (NYSTAR), Albany, NY 12245 http://www.esd.ny.gov/
North Carolina	North Carolina Manufacturing Extension Partnership 1005 Capability Drive, Research II Building., Suite 200, Raleigh, NC 27695 http://www.ncmep.org
North Dakota	North Dakota Manufacturing Extension Partnership 1929 North Washington Street, Suite M , Bismarck, ND 58501 http://www.impactdakota.com
Ohio	Ohio Manufacturing Extension Partnership 77 South High Street, Columbus, OH 43215 http://www.development.ohio.gov
Oklahoma	Oklahoma Manufacturing Alliance 525 South Main Street, Suite 210, Tulsa, OK 74103 http://www.okalliance.com/
Oregon	Oregon Manufacturing Extension Partnership 7650 SW Beveland Street,Suite 170, Portland, OR 97223 http://www.omep.org
Pennsylvania	Catalyst Connection 2000 Technology Drive, Pittsburgh, PA 15219 http://www.catalystconnection.org
	Delaware Valley Industrial Resource Center 2905 Southampton Road, Philadelphia, PA 19154 http://www.dvirc.org
	Innovative Manufacturers Center (IMC) One College Avenue, DIF 32, Williamsport, PA 17701 http://www.imcpa.com
	MANTEC 600 North Hartley Street, Suite 100, York,PA17404. http://www.mantec.org
	Manufacturers Resource Center 961 Marcon Boulevard, Suite 200, Allentown, PA 18109 http://www.mrcpa.org

Table A-1. (Continued)

State	Center Name, Address, and Website
	Northeastern Pennsylvania Industrial Resource Center 75 Young Street, Hanover Industrial Estates, Hanover Township, PA 18706 http://www.nepirc.com
	Northwest Pennsylvania Industrial Resource Center 5340 Fryling Road, Suite 202, Erie, PA 16510 http://www.nwirc.org
Puerto Rico	Puerto Rico Manufacturing Extension Inc. Ponce de Leon Avenue, Mercantil Plaza Building, Suite 819, Hato Rey, PR 00918 http://www.primexpr.org
Rhode Island	University of Rhode Island Research Foundation 75 Lower College Road, Suite 001, Kingston, RI 02881 http://www.polarismep.org
South Carolina	South Carolina Manufacturing Extension Partnership 250 Berryhill Road, Suite 512, Columbia, SC 29210 http://www.scmep.org
South Dakota	South Dakota Manufacturing and Technology Solutions 2329 N. Career Avenue,Suite 106, Sioux Falls, SD 57107 http://www.sdmanufacturing.com
Tennessee	Tennessee Manufacturing Extension Partnership 193 Polk Avenue, Ste.C, Univ.of Tennessee Center for Industrial Services, Nashville, TN 37210 http://www.cis.tennessee.edu/
Texas	TMAC 9390 Research Boulevard, SuiteII-300, Austin, TX 78759 http://www.tmac.org/
Utah	Utah Manufacturing Extension Partnership 815 West 1250 South, MS 212, Orem, UT 84058 http://www.mep.org/
Vermont	Vermont Manufacturing Extension Center P.O. Box 12, 1540 VT Rt.66,Suite103, Randolph, VT 05060 http://www.vmec.org/
Virginia	Genedge Alliance 32 Bridge Street, Suite 200, Martinsville, VA 24112 http://www.genedge.org
Washington	Impact Washington 8227 44thAvenue West, Suite D, Mukilteo, WA 98275 http://www.impactwashington.org
West Virginia	West Virginia Manufacturing Extension Partnership 886 Chestnut Ridge Road, 2ndFloor, Morgantown, WV 26506http://www.wvmep.com

State	Center Name, Address, and Website
Wisconsin	UW Stout Manufacturing Outreach Center 278 Jarvis Hall, 410 10thAve E, Menomonie, WI 54751 http://www.uwstout.edu/moc
	Wisconsin Manufacturing Extension Partnership 2601 Crossroads Drive, Suite 145, Madison, WI 53718 http://www.wmep.org
Wyoming	Manufacturing-Works Department 3362, 1000 East University Avenue, Laramie, WY 82071 http://www.manufacturing-works.com/

Source: Information provided by NIST to CRS via email, October 20, 2015.

APPENDIX B. CENTER FUNDING FOR ROUNDS ONE AND TWO OF THE SYSTEM-WIDE COMPETITION

Table B-1. First-Year Center Funding Awarded in Round One

(by state, in current dollars)

State	First Year NIST Funding
Colorado	$ 1,668,359
Connecticut	1,476,247
Indiana	2,758,688
Michigan	4,299,175
New Hampshire	628,176
North Carolina	3,036,183
Oregon	1,792,029
Tennessee	1,976,348
Texas	6,700,881
Virginia	1,722,571

Source: NIST, "NIST Awards $26 Million to Support Manufacturing in 10 States," press release, February 24, 2015, http://www.nist.gov/mep/awards-support-manufacturing.cfm.

Table B-2. First-Year Center Funding Awarded in Round Two

(by state, in current dollars)

State	First Year NIST Funding
Alaska	$ 500,000
Idaho	640,236
Illinois	5,029,910
Minnesota	2,653,649
New Jersey	2,814,432
New York	5,985,194
Oklahoma	1,309,080
Washington	2,534,872
West Virginia	500,000
Wisconsin	3,250,792

Source: NIST, "New Funding Brings New Opportunities for Manufacturers in Nine States," press release, September 21, 2015, http://www.nist.gov/mep/new-funding-brings-new-opportunities-for-manufacturers-in-ninestates.cfm; NIST, "New Funding Awarded to Support Wisconsin Manufacturers," press release, November 13, 2015, http://www.nist.gov/mep/new-funding-awarded-to-support-wisconsin-manufacturers.cfm.

Notes: NIST anticipates the awarding of the Utah center, competed in the second round, in early FY2016. No applications were received for the Ohio center which was also part of the second round competition.

End Notes

[1] NIST is an agency of the U.S. Department of Commerce.

[2] NIST defines SMMs as manufacturers with 500 or fewer employees.

[3] NIST website, Manufacturing Extension Partnership Strategic Plan, http://www.nist.gov /mep/about/strategicplan.cfm.

[4] National Institute of Standards and Technology, FY2016 Congressional Budget Justification, p. NIST-228, http://www.osec.doc.gov/bmi/budget/FY16CJ/NIST-NTIS_FY_2016_CJ_Final_508_Compliant.pdf.

[5] Telephone conversation between CRS and NIST, January 12, 2016.

[6] National Institute of Standards and Technology, FY2016 Congressional Budget Justification, p. NIST-229, http://www.osec.doc.gov/bmi/budget/FY16CJ/NIST-NTIS_FY_2016_CJ_Final_508_Compliant.pdf.

[7] Ibid., p. NIST-35.

[8] John Bulloch, "Accomodating the Future," Journal of Small Business and Entrepreneurship, vol. 5, no. 2 (Fall 1987), p. 8.

[9] Department of Commerce, Census Bureau, Statistics of U.S. Businesses Main, Table, http://www2.census.gov/econ/usb/data/2012/us_state_naicssector_small_emplsize_2012.xls

[10] National Institute of Standards and Technology, 2014 Economic Impact Survey: Making an Impact on U.S. Manufacturing, 2015, p. 2, http://www.nist.gov/mep/about/upload/MEP-ECONOMIC-IMPACTS-FY2014.pdf.

[11] Ibid., p. NIST-22.

[12] P.L. 100-418.

[13] P.L. 111-358.

[14] Dave Cranmer, Reflections—Part 2, Manufacturing Innovation blog, http://nistmep.blogs.govdelivery.com/reflections-part-2/.

[15] Dave Cranmer, Reflections—Part 1, Manufacturing Innovation blog, http://nistmep.blogs.govdelivery.com/25-yearreflections/.

[16] General Accounting Office, Technology Transfer, Federal Efforts to Enhance the Competitiveness of Small Manufacturers, GAO/RCED-92-30, November 1991, p. 3.

[17] General Accounting Office, Manufacturing Extension Program, Manufacturers' Views About Delivery and Impact of Services, GAO/GGD-96-75, March 1996, 2.

[18] Dave Cranmer, Reflections—Part 2, Manufacturing Innovation blog, http://nistmep.blogs.govdelivery.com/reflections-part-2/.

[19] Slides provided by Roger D. Kilmer, Director, Hollings Manufacturing Extension Partnership, NIST, May 19, 2010.

[20] Personal communication with MEP staff, October 8, 2015.

[21] National Institute of Standards and Technology, presentation, "Advisory Board Committee on Technology Acceleration (ABCTA) Report to the MEP Advisory Board," September 24, 2014.

[22] For more information on the SBIR program, see CRS Report R43695, Small Business Innovation Research and Small Business Technology Transfer Programs, by John F. Sargent Jr.

[23] For more information on the NNMI, see CRS Report R43857, The Network for Manufacturing Innovation, by John F. Sargent Jr.

[24] Personal communication with MEP staff, October 8, 2015.

[25] 15 USC 278k(a).

[26] According to NIST, the reimbursement structure for services varies among MEP centers. NIST MEP provides centers with flexibility in programmatic approaches and financial models, while requiring adherence to strict compliance with accounting systems, board governance, and reporting. NIST MEP does not provide MEP centers with guidance on charging clients. Source: email communication between NIST and CRS on November 22, 2015.

[27] Email communication between NIST and CRS, October 30, 2015.

[28] In OMB Circular A-11 (Preparation, Submission, and Execution of the Budget), the Office of Management and Budget defines full-time equivalent (FTE) employment as "the basic measure of the levels of employment used in the budget. It is the total number of hours worked (or to be worked) divided by the number of compensable hours applicable to each fiscal year."

Source: https://www.whitehouse.gov/sites/default/files/omb/assets/a11_current_year/a11_2015.pdf.

[29] National Institute of Standards and Technology, FY2016 Congressional Budget Justification, p. NIST-227, http://www.osec.doc.gov/bmi/budget/FY16CJ/NIST-NTIS_FY_2016_CJ_Final_508_Compliant.pdf.

[30] 15 USC 278k(e).

[31] The Advisory Board is exempted from the provisions of Section 14 of the Federal Advisory Committee Act, which addresses questions related to termination, renewal, and continuation of advisory committees.

[32] According to NIST, "The definition of a service location is broad in that it encompasses locations for which an MEP practitioner can operate out of in order to provide support for the manufacturing community. Service locations range from one-person offices to fully staffed regional offices with all service locations intended to provide adequate coverage for manufacturers. This includes partner locations that can be used to provide services to the manufacturers across the states." Source: Email communication between NIST and CRS, November 22, 2015.

[33] National Institute of Standards and Technology, FY2016 Congressional Budget Justification, p. NIST-227, http://www.osec.doc.gov/bmi/budget/FY16CJ/NIST-NTIS_FY_2016_CJ_Final_508_Compliant.pdf.

[34] 15 U.S.C. 278k(c)(4).

[35] Explanatory Statement, Consolidated and Further Continuing Appropriations Act, 2013.

[36] GAO referred to the imbalance as being inconsistent with the "beneficiary equity standard." According to GAO, "This standard—which is commonly used in social science research to design and evaluate funding formulas—calls for funds to be distributed in a way that takes these variations into account so that centers can provide the same level of services to each target manufacturing firm, according to its needs." Source: US Government Accountability Office, Most Federal Spending Directly Supports Work with Manufacturers, but Distribution Could Be Improved, GAO-14- 317, p. 15.

[37] U.S. Government Accountability Office, Most Federal Spending Directly Supports Work with Manufacturers, but Distribution Could Be Improved, GAO-14-317, March 2014, p. 15.

[38] Ibid, "GAO Highlights."

[39] Ibid, p. 20.

[40] Telephone conversation between NIST MEP and CRS, October 23, 2015.

[41] NIST, press release, "NIST Awards $26 Million to Support Manufacturing in 10 States," February 24, 2015, http://www.nist.gov/mep/awards-support-manufacturing.cfm.

[42] In the first round of the re-competition, NIST awarded new cooperative agreements to centers in Colorado, Connecticut, Indiana, Michigan, New Hampshire, North Carolina, Oregon, Tennessee, Texas, and Virginia. In the second round, NIST awarded new cooperative agreements to centers in Alaska, Idaho, Illinois, Minnesota, New Jersey, New York, Oklahoma, Washington, West Virginia, and Wisconsin. NIST anticipates the awarding of the Utah center, competed in the second round, in early FY2016.

[43] Email communication between NIST and CRS, October 23, 2015.

[44] Meeting between NIST MEP and CRS, October 8, 2015.

[45] Email communication between NIST MEP and CRS, October 23, 2015.

[46] Meeting between NIST MEP and CRS, October 8, 2015.

[47] Email communication between NIST MEP and CRS, October 23, 2015.

[48] NIST anticipates that the third round will include center competition for Alabama, Arkansas, California, Georgia, Louisiana, Massachusetts, Missouri, Montana, Pennsylvania, Puerto Rico, and Vermont, and the fourth round will include center competitions for Delaware, Hawaii, Iowa, Kansas, Maine, Mississippi, New Mexico, Nevada, North Dakota, South Carolina, and Wyoming. Source: NIST, "NIST Announces 2016 Plans for Manufacturing Center Competitions," press release, May 15, 2015, http://www.nist.gov/mep/2016-mep-comp-plans.cfm.

[49] National Institute of Standards and Technology, FY2016 Congressional Budget Justification, p. NIST-7, http://www.osec.doc.gov/bmi/budget/FY16CJ/NIST-NTIS_FY_2016_CJ_Final_508_Compliant.pdf.

[50] National Institute of Standards and Technology,, "Award Competitions for Hollings Manufacturing Extension Partnership (MEP)," 79 Federal Register 44746-44752, August 1, 2014, https://federalregister.gov/a/2014-18264.

[51] Email communication between NIST and CRS, slide presentation, October 30, 2015.

[52] 15 U.S.C. 278k(c)(3)(B). Also, according to NIST, "MEP Centers are not limited to working with SMMs, as the program is authorized to provide services to all manufacturers. Some MEP Centers work with non-manufacturers from time to time or as a separate business line through other partnership arrangements. Those costs cannot be used for reimbursement or matching on our award. [Centers] also cannot count any Federal funding they receive from other sources, including any funding sent through the state." Source: Email communication from NIST to CRS, November 23, 2015.

[53] 15 U.S.C. 278k(c)(5), subsequently amended by P.L. 105-309.

[54] In a 1995 study, the U.S. General Accounting Office found that firms that used internal funding to implement recommendations offered by extension programs were the most likely to find an overall positive impact on their manufacturing position Source: U.S. General Accounting Office, Manufacturing Extension Programs, Manufactures' Views of Service, GAO/GGD-95-216BR, August 1995.

[55] E.S. Oldsman, G.M. Ugiansky, and R. Jamin, Review of Mission and Operations of Regional Centers, National Institute of Standards and Technology, February 1, 1998, available at http://www.nist.gov/cgi-bin/view_pub.cgi? pub_id=200288&divison=260.

[56] P.L. 104-208 and P.L. 105-277, respectively.

[57] S.Rept. 100-80, p. 15.

[58] Ibid., p. 17.

[59] Slides provided by Roger D. Kilmer, Director, Hollings Manufacturing Extension Partnership, NIST, May 19, 2010.

[60] Government Accountability Office, Factors for Evaluating the Cost Share of Manufacturing Extension Partnership Program to Assist Small and Medium-Sized Manufacturers, GAO-11-437R, April 4, 2011, p. 4, http://www.gao.gov/assets/100/97395.pdf.

[61] Ibid., p. 4.

[62] CBO, Reducing the Deficit: Spending and Revenue Options, March 10, 2011, https://www.cbo.gov/sites/default/files/112th-congress-2011-2012/reports/03-10-reducing thedeficit.pdf. This issue is discussed in more detail later in the report. See "Congressional Budget Office," pp. 17-18.

[63] Both H.R. 5035 (113th Congress) and S. 2757 (113th Congress) defined "costs incurred" as costs incurred in connection with the activities undertaken to improve the competitiveness, management, productivity, and technological performance of small and medium-sized manufacturing companies.

[64] MEP Advisory Board, 2014 Annual Report, http://www.nist.gov/mep/about/upload/Advisory-Board-Annual-Report2014.pdf.

[65] S.Rept. 114-66.

[66] Only the centers in Arizona, Florida, Kentucky, Maryland, Nebraska, Rhode Island, and South Dakota were not in or past the fifth year of their cooperative agreements prior to the start of the system-wide competition. Source: Email communication between NIST and CRS, November 5, 2015.

[67] Funding for the B2B awards was provided via reprogramming of $2.5 million in FY2014 appropriations from the NIST Technology Innovation Program. Source: Letter from Ellen Herbst, Chief Financial Officer and Assistant Secretary for Administration, Department of Commerce, to Senator Barbara Mikulski, Chairwoman, Senate Committee on Appropriations, March 7, 2014.

[68] National Institute of Standards and Technology, FY2016 Congressional Budget Justification, pp. NIST-229-NIST230, http://www.osec.doc.gov/bmi/budget/FY16CJ/NIST-NTIS_FY_2016_CJ_Final_508_Compliant.pdf; NIST, press release, "NIST Awards $2.5 Million in Grants to MEP Centers for Pilot Business-to-Business Networks," December 2, 2014, http://www.nist.gov/mep/mep-120214.cfm.

[69] The award recipients were: Maine MEP; Michigan Manufacturing Technology Center; InnovateMEP Mississippi; Missouri Enterprise; Ohio MEP (State of Ohio, Ohio Development Services Agency: two awards, including the Appalachian Partnership for Economic Growth and the Manufacturing Advocacy and Growth Network); Oregon MEP; Northeastern Pennsylvania Industrial Resource Center; South Carolina MEP; and Impact Washington. Source: Email communication between NIST and CRS, November 5, 2015.

[70] Participating agencies include the NIST, the Department of Commerce's Economic Development Administration, the Department of Energy, the Department of Labor's Employment and Training Administration, the Small Business Administration, and the National Science Foundation.

[71] NIST, The Advanced Manufacturing Jobs and Innovation Accelerator Challenge (AMJIAC): Mid-Project Review, May 2014, http://www.nist.gov/mep/upload/AMJIAC-Report-final0520.pdf.

[72] NIST, Manufacturing Technology Acceleration Center (M-TAC) Pilot Project): Report on Initial Progress and Learning, February 2015, p. 5, http://www.nist.gov/mep/services/supplychain/upload/MTAC_Report-print.pdf.

[73] NIST, "NIST Manufacturing Extension Partnership Awards $9.1 Million for 22 Projects to Enhance U.S. Manufacturers' Global Competitiveness," press release, October 5, 2010, http://www.nist.gov/mep/mep_100510.cfm.

[74] Ibid.

[75] NIST, System Strategic Plan, Draft 11/07/14, November 7, 2014, http://www.nist.gov/mep/upload/STRATEGICplan-booklet-summary.pdf.

[76] Other comments and recommendations by these organizations are included elsewhere in this report.

[77] MEP Advisory Board, 2014 Annual Report, http://www.nist.gov/mep/about/upload/Advisory-Board-Annual-Report2014.pdf.

[78] MEP Advisory Board, 2014 Annual Report, http://www.nist.gov/mep/about/upload/Advisory-Board-Annual-Report2014.pdf.

[79] Government Accountability Office, Most Federal Spending Directly Supports Work with Manufacturers, but Distribution Could Be Improved, GAO-14-317, March 2014.

[80] America COMPETES Reauthorization Act (P.L. 111-358).

[81] Government Accountability Office, Factors for Evaluating the Cost Share of Manufacturing Extension Partnership Program to Assist Small and Medium-Sized Manufacturers, GAO-11-437R, April 4, 2011, p. 4, http://www.gao.gov/assets/100/97395.pdf.

[82] Government Accountability Office, Factors for Evaluating the Cost Share of Manufacturing Extension Partnership Program to Assist Small and Medium-Sized Manufacturers, GAO-11-437R, April 4, 2011, p. 4, http://www.gao.gov/assets/100/97395.pdf.

[83] U.S. Government Accountability Office, Manufacturing Extension Programs, Manufacturers' Views of Services, GGD-95-216BR, August 7, 1995, http://gao.gov/products/GGD-95-216BR.

[84] Government Accountability Office, Manufacturing Extension Programs, Manufacturers' Views about Delivery and Impact of Services, GGD-96-75, March 14, 1996, p. 3, http://gao.gov/products/GGD-96-75.

[85] Congressional Budget Office, Budget Options: Volume 2, 370-372, p. 88, August 2009, https://www.cbo.gov/sites/default/files/111th-congress-2009-2010/reports/08-06-budgetoptions.pdf.

[86] Ibid.

[87] National Academy of Public Administration, The Manufacturing Extension Partnership Program, Report 2, Alternative Business Models, May 2004, available at http://www.napawash.org/Pubs/NIST6-2-04.pdf.

[88] Manufacturing Extension Partnership, Next Generation Strategic Plan, 2006, http://www.mep.nist.gov/documents/pdf/about-mep/Next_Gen_MEP_Strategy.pdf.

[89] Email communication between NIST MEP and CRS, October 23, 2015.

[90] Commerce, Justice, Science, and Related Agencies Appropriations Act, 2016 (H.R. 2578).

[91] S.Rept. 114-66.

[92] NIST, Fiscal Year 2009 Budget Submission to Congress, http://www.osec.doc.gov/bmi/budget/09CBJ/NISTand%20NTIS%20FY2009%20Congressional%20Justification.pdf.

[93] The GDP (Chained) Price Index, a measure used by the Office of Management and Budget to adjust for inflation in research and development, grew at 1.8% CAGR during this period.

[94] Explanatory Statement, Consolidated and Further Continuing Appropriations Act, 2013, Congressional Record, March 11, 2013, p. 1301.

[95] Government Accountability Office, Most Federal Spending Directly Supports Work with Manufacturers, but Distribution Could Be Improved, GAO-14-317, March 2014.

In: The Manufacturing Extension Partnership … ISBN: 978-1-63485-193-0
Editor: Randolph S. Perkins

Chapter 2

MANUFACTURING EXTENSION PARTNERSHIP: MOST FEDERAL SPENDING DIRECTLY SUPPORTS WORK WITH MANUFACTURERS, BUT DISTRIBUTION COULD BE IMPROVED*

United States Government Accountability Office

ABBREVIATIONS

DVIRC	Delaware Valley Industrial Resource Center
E-CAR	Expansion of Services Cooperative Agreement Recipients
E3	economy, energy, and environment
GDP	gross domestic product
IT	information technology
MEP	Manufacturing Extension Partnership
NBS	National Bureau of Standards
NIST	National Institute of Standards and Technology
OMB	Office of Management and Budget
T-CAR	Tool Development Cooperative Agreement Recipients

* This is an edited, reformatted and augmented version of The United States Government Accountability Office publication, GAO-14-317, dated March 2014.

WHY GAO DID THIS STUDY

Manufacturing plays a key role in the U.S. economy. Congress established the MEP program in NIST in 1988. The program's objectives are to enhance productivity and technological performance and to strengthen the global competitiveness of target manufacturing firms, namely small and medium-sized U.S.-based firms. Under this program, NIST partners with 60 nonfederal organizations called MEP centers. The centers, located in 50 states and Puerto Rico, help target firms develop new customers and expand capacity, among other things. NIST awards federal funding to centers under annually renewed cooperative agreements, subject to the centers providing matching funds and receiving a positive performance evaluation.

The Consolidated and Further Continuing Appropriations Act, 2013, mandated GAO to report on MEP program administrative efficiency, which relates to funding provided to centers. This report (1) describes, over the past 5 years, how NIST spent federal MEP program funds and (2) examines the basis for NIST's cooperative agreement award spending. To conduct this work, GAO analyzed obligations data, reviewed relevant legislation and policies, and interviewed NIST officials.

WHAT GAO RECOMMENDS

GAO recommends that Commerce's spending on cooperative agreement awards be revised to account for variations across service areas in demand for program services and in MEP centers' costs of providing services. Commerce agreed with GAO's recommendation.

WHAT GAO FOUND

Of the approximately $608 million spent by the Department of Commerce's (Commerce) National Institute of Standards and Technology (NIST) in fiscal years 2009 through 2013 on the Manufacturing Extension Partnership (MEP) program, NIST used most of the funds to directly support MEP centers. Specifically, NIST spent about $495 million on awards to centers and spent the rest on contracts, staff, agency-wide overhead charges, and other items, some of which NIST considered direct support and some of which NIST considered administrative spending. Although NIST is not required to track, and has not historically tracked, administrative spending, NIST officials told GAO the agency developed definitions of direct support and administrative spending in fiscal year 2013 in response to congressional interest, then conducted an analysis of fiscal year 2013 federal MEP program spending using those definitions. NIST defines direct support spending as spending that directly supports the MEP center system's work with manufacturing firms, such as awards to centers or contracts to train MEP center staff on how to quickly assess innovative ideas for new products. NIST considers all other spending to be administrative, including spending on performance evaluations for MEP centers or on agency-wide overhead fees that pay for facilities operations and maintenance at the NIST campus. Using these definitions, NIST estimated that about 88.5 percent of federal MEP program spending in fiscal year 2013 was for direct support, and the remaining 11.5 percent was for administration.

NIST's spending on cooperative agreement awards is based on the historical amount awarded to each center when it was established. This took into account each center's identification of target manufacturing firms in its service area— including characteristics such as business size, industry types, product mix, and technology requirements—and its costs of providing services to those firms. However, because NIST made the awards on an incremental basis to individual centers serving different areas over a period of more than 15 years, NIST's awards did not take into account variations across service areas in the demand for program services—a function of the number and characteristics of target firms—or variations across service areas in costs of providing services. NIST's cooperative agreement award spending is, therefore, inconsistent with the beneficiary equity standard. This standard—commonly used to design and evaluate funding formulas—calls for funds to be distributed in a way that takes these variations into account so that centers can provide the same level of services to each target manufacturing firm, according to its needs. Because NIST did not account for these variations across service areas, NIST's cooperative agreement award spending may not allow centers to provide the same level of services to target manufacturing firms, according to their needs. NIST officials told GAO that an analysis they recently conducted showed wide variation across centers in the relationship between their cooperative agreement award amounts and the number of target manufacturing firms in their service areas. NIST officials told GAO they are exploring ways to revise NIST's cooperative agreement award spending to take into account variations across service areas in the number of target manufacturing firms, among other factors.

The officials discussed various options they are considering, but they did not identify an option they had agreed to implement or a timeline for decision making and implementation.

* * *

March 27, 2014

The Honorable Barbara Mikulski
Chairwoman

The Honorable Richard Shelby
Ranking Member
Subcommittee on Commerce, Justice, Science, and Related Agencies
Committee on Appropriations
United States Senate

The Honorable Frank Wolf
Chairman

The Honorable Chaka Fattah
Ranking Member
Subcommittee on Commerce, Justice, Science, and Related Agencies
Committee on Appropriations
House of Representatives

Manufacturing plays a key role in the U.S. economy. According to the Bureau of Economic Analysis, manufacturing generated $1.9 trillion in gross domestic product (GDP)—or almost 12 percent of total U.S. GDP— in 2012, the most recent year for which data are available. In addition, the manufacturing sector employed almost 12 million workers in July 2013 according to the Bureau of Labor Statistics. To support U.S. manufacturing, Congress has established a variety of tax preferences, subsidies, and other programs,[1] including the Hollings Manufacturing Extension Partnership (MEP) program. The MEP program was authorized by the Omnibus Trade and Competitiveness Act of 1988 and is administered by the Department of Commerce's National Institute of Standards and Technology (NIST).[2] The program's objectives are to enhance the productivity and technological performance and to strengthen the global competitiveness of target manufacturing firms, namely small and medium-sized manufacturing firms (with fewer than 500 employees) based in the United States, helping these firms create and retain jobs.[3]

The MEP program operates through a network of 60 nonfederal organizations—called MEP centers—located throughout the United States and in Puerto Rico. MEP centers enter into contracts with manufacturing firms under which the centers deliver technical assistance to improve the firms' manufacturing processes and productivity; expand their capacity; and help them adopt new technologies, utilize best management practices, and accelerate company growth. NIST enters into cooperative agreements with each MEP center under which federal funding is awarded to the center conditional upon the center providing matching funds and receiving a positive performance evaluation;[4] these agreements are renewed annually.[5] Once established, a MEP center must obtain two-thirds of its budget from nonfederal sources, such as state or local governments or user fees collected from manufacturing firms, to receive a one-third federal funding match.[6] (See fig. 1.) NIST has also awarded competitive grants to some MEP centers in recent years in addition to their annual cooperative agreement funding.

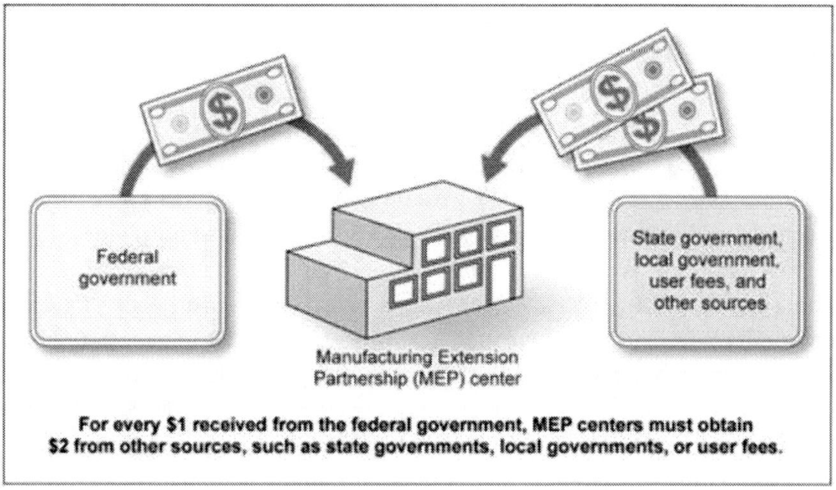

Source: GAO.

Figure 1. Manufacturing Extension Partnership Program Cost Share.

This engagement responds to a mandate in the explanatory statement accompanying the Consolidated and Further Continuing Appropriations Act, 2013 directing GAO to report to the House and Senate Appropriations

Committees, Subcommittees on Commerce, Justice, Science, and Related Agencies on the extent to which the MEP program achieves administrative efficiencies, which relates to funding provided to centers. Our objectives for this review were to (1) describe, over the past 5 years, how NIST spent federal MEP program funds and (2) examine the basis for NIST's cooperative agreement award spending.

To describe how NIST spent federal MEP program funds over the past 5 years, we obtained and analyzed obligations data from NIST for fiscal years 2009 through 2013. For the purposes of this report, we are measuring spending based on obligations. We interviewed knowledgeable agency officials about the sources of the data and the controls NIST had in place to maintain the integrity of the data, and we determined that the data were sufficiently reliable for the purposes of our report. We also reviewed a NIST analysis of fiscal year 2013 spending that estimated the amount NIST spent for direct support of the MEP center system's work with manufacturing firms and the amount NIST spent for administration. We interviewed agency officials about the analysis and independently reviewed their determinations of administrative and direct support spending, and we determined the resulting estimates were sufficiently reliable for the purposes of our report. In addition, we reviewed relevant laws—including the MEP program's enabling legislation, implementing regulations, and annual appropriations—and program policies and procedures concerning how NIST spends the federal program funds. We also reviewed relevant documents, such as strategic plans, performance reports, budget justifications, and excerpts of contracts, and we interviewed NIST officials about how they spent the funds. To examine the basis for NIST's cooperative agreement award spending, we interviewed NIST officials and reviewed MEP program data for fiscal years 2009 through 2013 to determine how NIST establishes cooperative agreement awards. We reviewed all federal funding opportunity announcements for the establishment of MEP center cooperative agreements published since the program's inception, which define the criteria used to establish cooperative agreement awards. We assessed NIST's basis for cooperative agreement award spending against equity standards from social science research for designing and evaluating funding formulas.

We conducted this performance audit from July 2013 to March 2014 in accordance with generally accepted government auditing standards. Those standards require that we plan and perform the audit to obtain sufficient, appropriate evidence to provide a reasonable basis for our findings and conclusions based on our audit objectives. We believe that the evidence obtained provides a reasonable basis for our findings and conclusions based on our audit objectives.

BACKGROUND

The MEP program traces its origins to the Manufacturing Technology Centers Program,[7] which was established by NIST's predecessor, the National Bureau of Standards (NBS). In July 1988, NBS published the first of 12 federal funding opportunity announcements that have been issued to date for the establishment of MEP centers.[8] That announcement led to the establishment of the first 3 centers in 1989, as part of an initial pilot program. By 1990, NBS had become NIST, and the agency published the second federal funding opportunity announcement for the establishment of two additional centers, bringing the total number of centers to 5. In 1992, NIST announced a federal funding opportunity for 2 more centers, bringing the total to 7. The number of centers grew rapidly thereafter, with a nationwide network of 44 centers in place by 1995. NIST has since added to its network and, as of 2013, has 60 centers that cover all 50 states and Puerto Rico. Appendix I provides a list of the 60 MEP centers.

The original legislation authorizing the MEP program emphasized the transfer of advanced technologies developed within NIST and other federal laboratories to small and medium-sized manufacturing firms. As we reported in 1991, however, the centers soon found that firms primarily needed proven, not advanced, technologies because advanced technologies were generally expensive, untested, and too complex to be practical for most small manufacturing firms. [9] We reported, therefore, that a key mandate of the program was not realistically aligned with the basic needs of most small manufacturing firms. In recognition of this situation, NIST reoriented the program to focus on basic technologies that permitted firms to improve their competitive position. By the time we reported on the program in 1996, centers were providing a wide range of business services, including helping companies solve individual manufacturing problems, obtain training for their workers, create marketing plans, and upgrade their equipment and computers.[10]

More recently, the program's focus has evolved to an emphasis on enhancing manufacturing firms' innovative capacity. NIST calls this its Next Generation Strategy and, in December 2008, NIST released its current strategic plan referred to by that name. The plan articulates NIST's new vision for the program as a catalyst for accelerating manufacturing's transformation into a "more efficient and powerful engine of innovation driving economic growth and job creation."[11] The plan also defines the program's mission: "to act as a strategic advisor to promote business growth and connect manufacturing firms to public and private resources essential for increased competitiveness and profitability." The plan focuses the program's activities around the following five strategic areas:

- *Continuous Improvement.* This area includes enhancing manufacturing firms' productivity and freeing up their capacity to provide them a stable foundation to pursue innovation and growth through services and programs that target manufacturing plant efficiencies.
- *Technology Acceleration.* This area includes developing tools and services to bring new product and process improvement opportunities to manufacturing firms, accelerating firms' opportunities to leverage and adopt technology, connecting firms with technology opportunities and solutions, and making available a range of product development and commercialization assistance services.
- *Supplier Development.* This area includes developing and delivering the national capacity, tools, and services needed to put suppliers in a position to thrive in existing and future global supply chains.
- *Sustainability.* This area includes helping companies gain a competitive edge by reducing environmental costs and impact by developing new environmentally focused materials, products, and processes to gain entry into new markets.
- *Workforce.* This area includes developing and delivering training and workforce assistance to manufacturing firms, as well as expanding partnerships and collaborations to develop and deliver tools and services to foster the development of progressive managers and entrepreneurial CEOs.

MEP centers work with manufacturing firms to plan and implement projects in these and other areas. For example, in 2011, the Delaware Valley Industrial Resource Center (DVIRC)—a MEP center in Pennsylvania—worked with a manufacturing firm in Hatfield, Pennsylvania, on a continuous improvement project when the company faced price increases from its vendors it did not want to pass on to its customers. DVIRC trained company staff on methods to achieve efficiencies and helped identify areas for improvement in the company's production process, resulting in increased productivity and reduced inventory levels that allowed the company to save space and lower costs, as reported by the manufacturing firm. Similarly, the Texas MEP center worked with a manufacturing firm in El Paso, Texas, on a sustainability project in 2013. The MEP center and the firm partnered with New Mexico State University's Institute for Energy and the Environment on an economy, energy, and environment (E3) project that included training and an effort to identify inefficiencies in the manufacturing process. As a result of this partnership, the firm reported saving 40,000 gallons of water and reducing solid waste by 56 tons, among other accomplishments.

NIST has recently begun developing a new strategic planning process for the MEP program to update its Next Generation Strategy. According to NIST officials, the process will include extensive participation by stakeholders, including MEP centers. NIST expects to implement the planning process through the spring of 2014 and release an updated strategic plan shortly after the planning process is complete.

The program has also evolved in its matching fund requirements. The program as originally implemented provided federal funding to reimburse each $1 of nonfederal contributions with no more than $1 of federal funding—referred to as a 1:1 cost share—for the first 3 years that a center operated. For the fourth year of operation, every $3 of nonfederal contributions were reimbursed with $2 of federal funding—referred to as a 3:2 cost share. For the fifth and sixth years of operation, every $2 of nonfederal contributions were reimbursed with $1 of federal funding— referred to as a 2:1 cost share. Under the original legislation, federal funding was scheduled to end once a center had operated for 6 years. The 6-year federal funding limit was temporarily suspended by the fiscal year 1997 and 1998 appropriations acts and was eliminated in 1998 when Congress passed legislation[12] changing the program to, among other things, provide for continued federal funding and set the cost share at 2:1 for all centers that had been in operation for at least 6 years.[13]

Most of the $608.3 Million NIST Spent on the MEP Program in Fiscal Years 2009 through 2013 Directly Supported MEP Centers

NIST spent $608.3 million in federal funding on the MEP program in fiscal years 2009 through 2013 and used most of these funds to directly support MEP centers and their work with manufacturing firms. Specifically, NIST spent $494.6 million on cooperative agreement awards and competitive grant awards to MEP centers, which NIST considers to be direct support. NIST spent $78 million for contracts, and for NIST staff salaries and benefits, some of which NIST considers direct support and some administrative spending. The remaining $35.7 million was spent for agency-wide overhead charges and travel, training, performance awards, and other items, all of which NIST considers administrative spending. NIST defines direct support as spending that directly supports the MEP center system's work with manufacturing firms, such as awards to centers or spending on training for MEP center staff. NIST considers all other spending to be administrative, including spending on performance evaluations of centers and on agency-wide overhead charges that pay for facilities operations and maintenance at the NIST campus.

NIST is not required to track, and historically has not tracked, administrative spending, but NIST officials told us the agency developed its definitions of direct support and administrative spending in fiscal year 2013 in response to congressional interest. It then conducted an analysis of fiscal year 2013 federal MEP program spending using those definitions. NIST estimated that about 88.5 percent of federal MEP program spending in fiscal year 2013 was for direct support, and the remaining 11.5 percent was administrative. (See fig. 2.)

It is not possible to determine whether NIST's amount of administrative spending is appropriate because there is no standard definition of administrative spending for federal programs. In addition, conducting the analysis using different definitions could produce different results.

Defining Administrative Expenses

There is no standard definition of administrative expenses for federal programs. Executive Order 12837, issued on February 10, 1993, called on the Director of the Office of Management and Budget (OMB) to establish a definition of administrative expenses for agencies, but OMB did not develop a definition. Definitions and reporting of administrative expenses vary across public and private entities depending on their mission, priorities, services, clients, and on the purposes for which management needs the information.

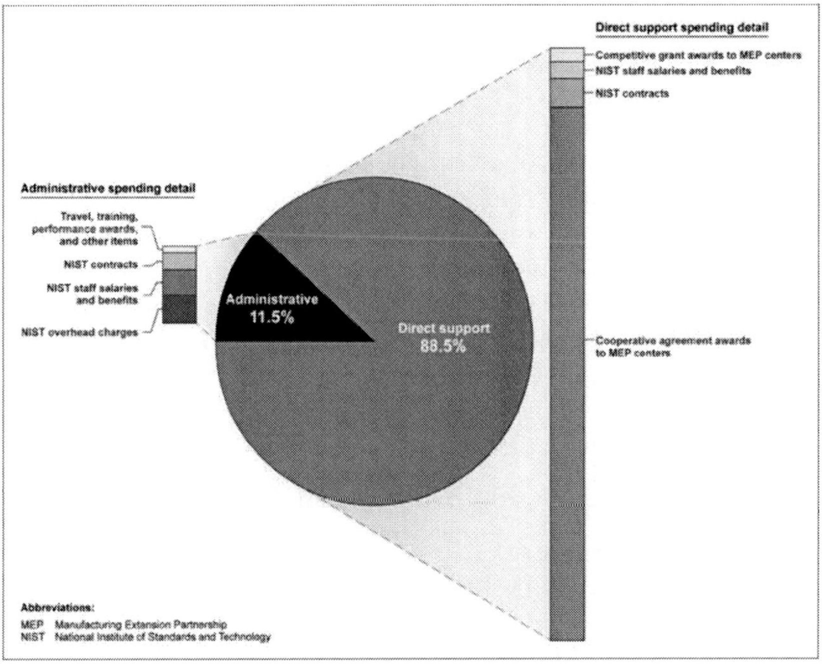

Source: GAO analysis of NIST data.

Figure 2. Manufacturing Extension Partnership Program Spending, Fiscal Year 2013.

Spending on MEP Center Cooperative Agreements

NIST spent $471 million on MEP center cooperative agreement awards in fiscal years 2009 through 2013. NIST considers all of this spending direct support. Federal funds for center cooperative agreements can be used by MEP centers for capital and operating and maintenance expenses. As stated earlier, these funds are awarded contingent on each MEP center meeting its cost-share requirement and having positive performance evaluations. NIST officials told us that spending on cooperative agreements and spending on staff salaries and benefits are NIST's top spending priorities.

Spending on MEP Center Competitive Grants

In fiscal years 2009 through 2013, NIST spent $23.6 million on awards to centers that were granted on a competitive basis. These awards are made in addition to cooperative agreement awards. The bulk of these awards went to two competitive grant programs; about $12.7 million was awarded to MEP centers through the Expansion of Services Cooperative Agreement Recipients (E-CAR) competition, and about $7.3 million was awarded through the Tool Development Cooperative Agreement Recipients (T-CAR) competition. NIST also awarded a small amount through grant competitions conducted under two other programs: the Advanced Manufacturing Jobs and Innovation Accelerator Challenge, and the Energy-Efficient Buildings Hub project.[14] According to NIST officials, the competitive grants that it awarded under the four programs encourage projccts in the five strategic areas identified in the MEP program's strategic plan. For example, NIST awards to MEP centers through the E-CAR competition funded 14 projects designed to integrate two or more of the five strategic areas, and NIST awards through the T-CAR competition funded 8 projects aimed at addressing the new and emerging needs of manufacturing firms in any of the strategic areas.

Spending on Contracts

In fiscal years 2009 through 2013, NIST spent $45.9 million on contracts for goods or services, some of which directly supported MEP centers and some of which were administrative. Of the $8 million spent on such contracts in fiscal year 2013, NIST estimated that it spent $5 million on direct support contracts and $3 million on administrative contracts. The contracts that NIST considered direct support in fiscal year 2013 were for training and support for MEP centers on tools that centers could use to assist manufacturing firms, or for work with centers on implementing MEP initiatives. For example, NIST awarded a $3.7 million contract to International Management and Consulting, LLC, to provide training for MEP center staff on innovation engineering, which NIST describes as a business support service that informs companies how to quickly assess innovative ideas resulting in new business models, processes, and products. The training is intended to help MEP center staff provide guidance in these areas to clients. Contracts that NIST considered administrative in fiscal year 2013 were for services related to performance evaluations of MEP centers, telephone and mobile broadband, information technology (IT), and other products and services.

According to NIST officials, some of these contracts helped the program meet legal requirements, such as for services related to performance evaluations of MEP centers, which are required by the program's enabling legislation and implementing regulations. Other contracts supported areas of the program's strategic plan, such as the contracts for MEP center staff training. Finally, some of the contracts were for operational functions, such as the contracts for telephone and mobile broadband and IT. NIST staff told us the program is currently reviewing all large direct support contracts with the intent of reducing contract spending and directing more funds to MEP centers. They expect the review to be complete in spring 2014.

Spending on Staff Salaries and Benefits

NIST spent $32.1 million on staff salaries and benefits in fiscal years 2009 through 2013. As of fiscal year 2013, NIST employed 55 staff under the MEP program. According to NIST's definitions, some of its staff directly supported MEP centers, and some were administrative. In fiscal year 2013, NIST estimated that it spent $2.9 million on direct support staff, and $4.4 million on administrative staff. As shown in figure 3, the direct support staff worked in NIST's strategic partnerships team, as well as in its program development and system operations offices for the MEP program. In NIST's 2013 analysis of administrative spending, NIST considered the strategic partnerships and program development staff to be entirely dedicated to directly supporting MEP centers, and the system operations staff to be half dedicated to directly supporting MEP centers and half dedicated to program administration. NIST considered the other six units and teams in the MEP program to be dedicated to program administration, including the Director and the administration and finance team. NIST officials told us that they increased spending for staff salaries and benefits during the past 5 years, in part to return the program's staffing level to that before substantial budget cuts in fiscal year 2004. New hiring focused on staff with expertise in areas of the program's strategic plan, according to these officials.

Spending on Agency-Wide Overhead

NIST spent $30.6 million in federal MEP program funds in fiscal years 2009 through 2013 on agency-wide overhead required by NIST. NIST considers this spending to be administrative. NIST does not receive an appropriation for the costs of agency-wide general administration; instead it levies surcharges on programs to pay overhead costs, including the operation and maintenance of facilities, grants management, and mail distribution.

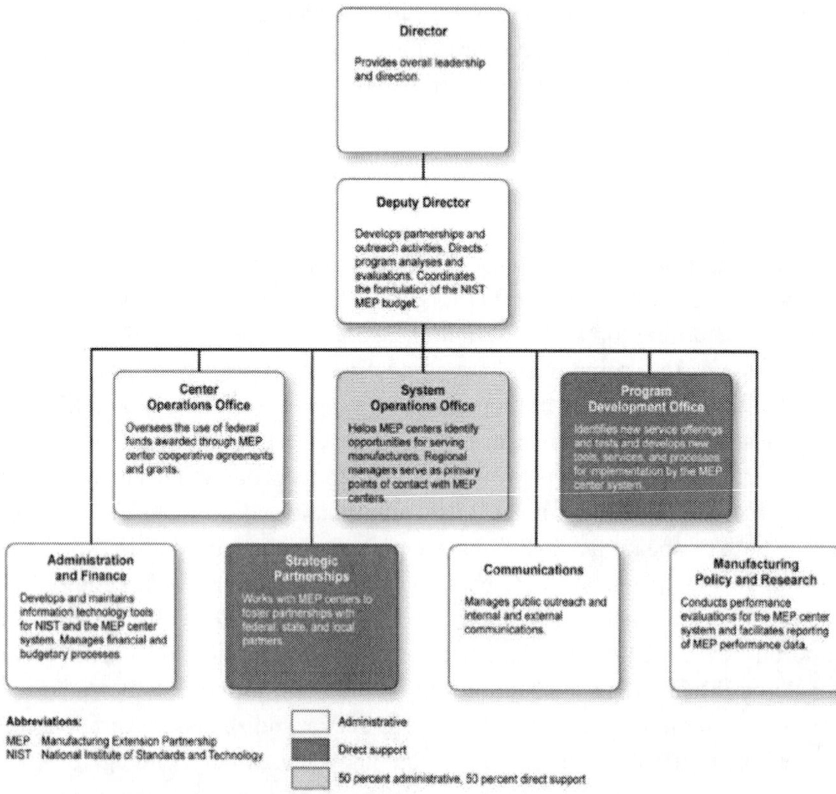

Source: GAO analysis of NIST information.

Figure 3. Manufacturing Extension Partnership Program Organizational Chart Showing Direct Support and Administrative Staff.

Spending on Travel, Training, Performance Awards, and Other Items

NIST spent the remainder of its federal funds—$5.1 million in fiscal years 2009 through 2013—on travel, training, staff performance awards, and other items. NIST considers all of this spending to be administrative. According to NIST's travel tracking spreadsheet, fiscal year 2013 travel included participation in on-site panel reviews of MEP centers, attendance at MEP center board meetings, and attendance at meetings with state and federal partners, among other things. Officials told us training funds are used for continuing education and professional training for MEP program staff, as opposed to training for MEP center staff. For example, some program staff hold professional credentials, such as a Contracting Officer Technical Representative, which require periodic training to be maintained. Performance award spending is used for discretionary bonuses and cash awards paid to MEP program staff for performance and to NIST staff outside the program, such as legal counsel, for exemplary support of the program.

Table 1 summarizes the spending described above.

Table 1. Manufacturing Extension Partnership (MEP) Program Spending, Fiscal Years 2009-2013

Dollars in millions		
Category	*Spending*	*Direct support or administrative spending*
Cooperative agreement awards to MEP centers	$471.0	Direct support
Competitive grant awards to MEP centers	23.6	Direct support
NIST contracts	45.9	Mix of direct support and administrative
NIST staff salaries and benefits	32.1	Mix of direct support and administrative
NIST overhead charges	30.6	Administrative
Travel, training, performance awards, and other items	5.1	Administrative
Total	*$608.3*	

Source: GAO analysis of NIST data.

NIST's Cooperative Agreement Award Spending Is Inconsistent with Beneficiary Equity

NIST's spending on cooperative agreement awards is based on the historical amount awarded to each center when it was established. This took into account each center's identification of target manufacturing firms in its service area—including characteristics such as business size, industry types, product mix, and technology requirements—and its costs of providing services to those manufacturing firms. However, because NIST made the awards on an incremental basis to individual centers serving different areas over a period of more than 15 years, NIST's awards to individual centers did not take into account variations across different service areas in the demand for program services—a function of the number and characteristics of target manufacturing firms—or variations across different service areas in costs of providing services. NIST's cooperative agreement award spending is, therefore, inconsistent with the beneficiary equity standard.[15] This standard—which is commonly used in social science research to design and evaluate funding formulas—calls for funds to be distributed in a way that takes these variations into account so that centers can provide the same level of services to each target manufacturing firm, according to its needs. Because NIST did not account for these variations across service areas, NIST's cooperative agreement award spending may not allow centers to provide the same level of services to target manufacturing firms, according to their needs. NIST officials told us that an analysis they recently conducted showed a wide variation across centers in the relationship between their cooperative agreement award amounts and the number of target manufacturing firms in their service areas. NIST officials told us they are exploring ways to revise NIST's cooperative agreement award spending to take into account variations across service areas in the number of target manufacturing firms, among other factors.

NIST's Cooperative Agreement Award Spending Is Based on Historical Practice

NIST's spending on cooperative agreement awards is based on the historical amount awarded to each center when it was established. Most of the currently operating centers were established between 1989 and 1996, according to our analysis of NIST data estimating the establishment dates of current centers.[16] When the centers were established, their original award amounts were based on the proposals that they submitted in response to NIST's federal funding opportunity announcements. For all but the first federal funding opportunity announcement, NIST specified that it would evaluate proposals by assigning scores to the following equally weighted criteria:[17]

- *Identification of target firms in the proposed region.* The proposals had to demonstrate an understanding of the service area's manufacturing base, including concentration of industry, business size, industry types, product mix, and technology requirements, among other things.
- *Technology resources.* The proposals had to assure strength in technical personnel and programmatic resources, full-time staff, facilities, equipment, and linkages to external sources of technology, among other things.
- *Technology delivery mechanisms.* The proposals had to define an effective methodology for delivering advanced manufacturing technology to manufacturing firms, among other things.
- *Management and financial plan.* The proposals had to define a management structure and assure management personnel to carry out development and operation of an effective center, among other things.
- *Budget.* The proposals had to contain a detailed 1-year budget and budget outline for subsequent years, among other things.

For funding opportunity announcements that NIST published after it issued its 2008 strategic plan, these criteria were to be discussed in the context of the proposer's ability to align the proposal with the program's strategic objectives. The announcements stated that, after scoring the proposals, NIST would select award recipients based upon their score ranking and other factors such as availability of federal funds and the need to assure appropriate regional distribution.

After centers were established, their subsequent cooperative agreement awards have remained at the historical amount when they were renewed each year.[18] According to NIST officials, in some instances, centers' cooperative agreements are not renewed and are instead opened to recompetition; during fiscal years 2009 to 2013, eight cooperative agreements were opened to recompetition.[19] NIST officials told us that recompetitions typically occur because the existing center has voluntarily closed or the organization has decided its mission no longer supports running a MEP center. According to NIST's funding opportunity announcements, NIST used the same evaluation criteria discussed above to select new centers and establish their awards. Unlike renewed cooperative agreement awards, which remain at the historical amount each year, recompeted awards are based on, but can be greater than, the historical amount. NIST officials told us that they use the historical amounts as a baseline in establishing the recompeted award amounts, but they may make additional funding available for the recompetitions. This was the case for all but one of the eight recompetitions that took place during fiscal years 2009 to 2013. According to NIST officials, during these years, NIST reserved additional funds for seven of the recompetitions to accommodate compelling proposals such as those that identified increased matching funds or broadened the work historically done in the service area. All but one of those seven recompetitions led to award amounts greater than the historical amount.

In addition to renewing existing awards and recompeting awards when a center has closed, NIST officials told us that NIST recently added a new center to the nationwide system and based the new award on the historical amount awarded for the area. Specifically, in 2012, a new center was added in South Dakota. Previously, the MEP center located in North Dakota served both North and South Dakota and received separate cooperative agreement awards for each. Serving both states proved to be difficult for the center, however, and most of its activity was focused in North Dakota. According to NIST officials, the state of South Dakota suggested to NIST the addition of a new South Dakota center. Through a competitive process, the new South Dakota center received an award equal in amount to the award that the North Dakota center previously received to serve South Dakota. The North Dakota center received an award equal in amount to the award it previously received to serve North Dakota.

NIST's Cooperative Agreement Award Spending Does Not Take into Account Variations in Demand for Program Services or in Centers' Costs of Providing Services across Service Areas

NIST's spending on cooperative agreement awards to MEP centers does not account for variations across centers' service areas in terms of the demand for program services, which is a function of the number and characteristics of target manufacturing firms. As a result, NIST's cooperative agreement award spending falls short of a component of beneficiary equity—a standard commonly used to design and evaluate funding formulas— that calls for funds to be distributed in a way that takes into account these variations so that each center can provide the same level of services to each target manufacturing firm, according to its needs. The original awards were made in part on the basis of each center's identification of target manufacturing firms in its service area, including characteristics such as business size, industry types, product mix, and technology requirements, among other things. NIST's funding opportunity announcements published in June 1995, May 1996, July 2000, March 2001, and March 2003 specified that award amounts should be directly related to the level of activity of the center, which is a function of the number of manufacturing firms in the designated service area. Because most of the current MEP center cooperative agreements were made on an incremental basis over a period of more than 15 years, they did not take into account the distribution of demand for program services across service areas. NIST officials told us they recognize that, as a result of the incremental addition of centers, wide variations emerged across centers in the relationship between their cooperative agreement award amounts and the number of target manufacturing firms in their service areas. Specifically, NIST officials told us that an analysis they recently conducted of current cooperative agreement award amounts per target manufacturing firm across service areas showed a mean of $333 per target manufacturing firm and a range of $82 to $972, with 75 percent of centers falling between $179 and $487. As a result, centers may not be able to provide the same level of services to each target manufacturing firm, according to its needs.

NIST's spending on cooperative agreement awards also does not take into account variations in MEP centers' costs of providing services to target manufacturing firms. As a result, NIST's cooperative agreement award spending falls short of another component of beneficiary equity. Under the beneficiary equity standard, funds should be distributed in a way that accounts for variations in the cost of providing services in each area, so that target manufacturing firms across MEP center service areas may receive the same level of assistance, according to their needs. The costs of operating the centers to provide assistance to manufacturing firms affect the amount of funding that centers have available for direct assistance to firms. According to NIST's funding opportunity announcements, costs—as presented by the centers' budgets—were considered in making the original awards, but these costs were presented on an incremental basis over a period of more than 15 years and, therefore, NIST's consideration of these costs did not account for variations across service areas. By not accounting for these variations, NIST's cooperative agreement award spending may further call into question centers' ability to provide the same level of services to each target manufacturing firm, according to its needs.

NIST Is Undertaking Efforts to Revise Its Cooperative Agreement Award Spending

NIST officials told us they are exploring ways to revise cooperative agreement award spending to take into account variations across service areas in the number of target manufacturing firms, among other factors. The officials discussed various options they are considering, but they did not identify an option they had agreed to implement or a timeline for decision making and implementation. They stated that one option they are considering is to provide increased awards to those centers that are currently underfunded relative to the mean relationship between centers' cooperative agreement award amounts and the number of manufacturing firms in their service areas. They told us that doing so would result in a greater benefit in terms of manufacturing firms served than providing additional funds to centers that are overfunded relative to the mean. NIST estimates that if it were to increase cooperative agreement award amounts for the underfunded centers, the program would see up to a 20 percent increase in the number of manufacturing firms served in these service areas over a 3-year period.

NIST officials told us that they face at least two impediments in revising cooperative agreement award spending. First, they stated that revising cooperative agreement award spending within the current level of funding would likely mean taking funds from some centers to give to others, and NIST is concerned about the effect this disruption might have on the impact of the program. The officials told us that they would like to increase NIST's total cooperative agreement award spending and that they are exploring options to do so. They said they are examining their spending on direct support contracts to determine whether cost savings can be realized and redirected to cooperative agreement awards. They also said that they are considering making any changes over a multiyear period. Our prior work has shown that phasing in changes to funding levels gradually over a number of years minimizes disruptions to funding recipients by providing them time to adjust.[20]

The second impediment that the officials identified is the requirement in the MEP program's authorizing legislation that federal cooperative agreement funds provided to MEP centers after their sixth year of operation not exceed one-third of their capital and annual operating and maintenance costs. This requirement leaves the centers responsible for raising the remaining two-thirds of matching funds from other sources. NIST officials told us that many centers already face difficulties raising the required two-thirds of matching funds and may not be able to raise the additional funds needed to access an increased cooperative agreement award.

CONCLUSION

Manufacturing plays a key role in the U.S. economy, and NIST has established a nationwide system of MEP centers dedicated to supporting and strengthening the U.S. manufacturing base. However, because NIST's cooperative agreement award spending does not take into account variations across service areas in the demand for program services—a function of the number and characteristics of target manufacturing firms—or variations in MEP centers' costs of providing services, centers may not be able to provide the same level of services to each target manufacturing firm, according to its needs. NIST officials told us they are exploring ways to revise cooperative agreement award spending to take into account variations across service areas in the number of target manufacturing firms, among other factors. Revising NIST's cooperative agreement award spending poses challenges because it could result in award decreases for some centers, along with increases for others. However, there are ways to ease the transition, such as phasing in changes gradually to minimize disruption to centers and the manufacturing firms they serve.

RECOMMENDATION FOR EXECUTIVE ACTION

To ensure that NIST's spending on cooperative agreement awards to MEP centers is more equitable to manufacturing firms in different service areas, we recommend that the Secretary of Commerce revise the program's cooperative agreement award spending to account for variations across service areas in:

- the demand for program services—a function of the number and characteristics of target manufacturing firms—and

- MEP centers' costs of providing services.

Agency Comments

We provided a draft of this report to the Department of Commerce's NIST for review and comment. In its written comments, NIST generally agreed with our findings and recommendation. In commenting on our recommendation, NIST stated that information in our report could help NIST continue to efficiently operate the MEP program.

NIST also provided technical comments, which we incorporated into the report as appropriate.

Frank Rusco
Director, Natural Resources and Environment

Appendix I: List of Manufacturing Extension Partnership Centers, 2013

State	City	Name
AK	Anchorage	Alaska Manufacturing Extension Partnership[a]
AL	Birmingham	Alabama Technology Network
AR	Little Rock	Arkansas Manufacturing Solutions
AZ	Phoenix	Arizona Commerce Authority
CA	Torrance	California Manufacturing Technology Consulting
CA	San Ramon	Corporation for Manufacturing Excellence
CO	Denver	Colorado Association for Manufacturing and Technology
CT	Rocky Hill	Connecticut State Technology Extension Program
DE	Newark	Delaware Manufacturing Extension Partnership
FL	Celebration	Florida Manufacturing Extension Partnership[b]
GA	Atlanta	Georgia Manufacturing Extension Partnership
HI	Honolulu	INNOVATE Hawaii
IA	Ames	Iowa Center for Industrial Research and Services
ID	Boise	Idaho TechHelp
IL	Park Ridge	Illinois Manufacturing Excellence Center-Chicago

State	City	Name
		Region
IL	Peoria	Illinois Manufacturing Excellence Center-Downstate
IN	Indianapolis	Purdue Technical Assistance Program
KS	Overland Park	Mid-America Manufacturing Technology Center
KY	Bowling Green	Advantage Kentucky Alliance
LA	Lafayette	Manufacturing Extension Partnership of Louisiana
MA	Worcester	Massachusetts Manufacturing Extension Partnership
MD	Columbia	Maryland MEP[c]
ME	Augusta	Maine Manufacturing Extension Partnership
MI	Plymouth	Michigan Manufacturing Technology Center
MN	Minneapolis	Enterprise Minnesota
MO	Rolla	Missouri Enterprise
MS	Ridgeland	InnovateMEP Mississippi
MT	Bozeman	Montana Manufacturing Extension Center
NC	Raleigh	North Carolina Industrial Extension Service
ND	Bismarck	North Dakota Manufacturing Extension Partnership
NE	Lincoln	Nebraska Manufacturing Extension Partnership
NH	Concord	New Hampshire Manufacturing Extension Partnership
NJ	Morris Plains	New Jersey Manufacturing Extension Partnership
NM	Albuquerque	New Mexico Manufacturing Extension Partnership
NV	Reno	Nevada Industry Excellence
NY	Albany	Empire State Development's Division of Science, Technology and Innovation (NYSTAR)
OH	Columbus	Ohio Manufacturing Extension Partnership
OK	Tulsa	Oklahoma Manufacturing Alliance
OR	Portland	Oregon Manufacturing Extension Partnership
PA	Pittsburgh	Catalyst Connection
PA	Philadelphia	Delaware Valley Industrial Resource Center
PA	Williamsport	Innovative Manufacturers' Center
PA	York	MANTEC
PA	Bethlehem	Manufacturers Resource Center
PA	Hanover Township	Northeastern Pennsylvania Industrial Resource Center
PA	Erie	Northwest Pennsylvania Industrial Resource Center
PR	Hato Rey	Puerto Rico Manufacturing Extension
RI	Providence	Rhode Island Manufacturing Extension Services[d]

Appendix I. (Continued)

State	City	Name
SC	Columbia	South Carolina Manufacturing Extension Partnership
SD	Sioux Falls	South Dakota Manufacturing and Technology Solutions[e]
TN	Nashville	University of Tennessee Center for Industrial Services
TX	Austin	Texas Manufacturing Assistance Center
UT	Orem	Utah Manufacturing Extension Partnership
VA	Martinsville	GENEDGE ALLIANCE
VT	Randolph Center	Vermont Manufacturing Extension Center
WA	Mukilteo	Impact Washington
WI	Menomie	Northwest Wisconsin Manufacturing Outreach Center
WI	Madison	Wisconsin Manufacturing Extension Partnership
WV	Morgantown	West Virginia Manufacturing Extension Partnership
WY	Laramie	Manufacturing-Works

Source: GAO analysis of National Institute of Standards and Technology (NIST) information.

[a] According to NIST officials, the Alaska Manufacturing Extension Partnership closed in September 2013.

[b] According to NIST officials, the Florida Manufacturing Extension Partnership is expected to close by March 2014.

[c] According to NIST officials, the Maryland MEP was established in July 2013.

[d] According to NIST officials, Rhode Island Manufacturing Extension Services was established in February 2013.

[e] According to NIST officials, South Dakota Manufacturing and Technology Solutions was established in January 2013.

End Notes

[1] See GAO, Global Manufacturing: Foreign Government Programs Differ in Some Key Respects from Those in the United States, GAO-13-365 (Washington, D.C.: July 25, 2013).

[2] Pub. L. No. 100-418 § 5121(a) (Aug. 23, 1988)(codified at 15 U.S.C. § 278k).

[3] For a comparison of the MEP program to the Trade Adjustment Assistance for Firms program, see GAO, Trade Adjustment Assistance: Commerce Program Has Helped Manufacturing and Services Firms, but Measures, Data, and Funding Formula Could Improve, GAO-12-930 (Washington, D.C.: Sept. 13, 2012), 47-49.

[4] According to the 2013 General Terms and Conditions of the program, NIST is to conduct a performance evaluation annually that focuses on each center's alignment with the program's overall objectives, resource expenditures, activities planned for the next year, and any proposed changes to the center's budget, among other things. The results of this evaluation are to be incorporated into the center's project plan or budget. In every other year this evaluation takes the form of an independent Merit Review Panel, in which private experts—not connected with the center being reviewed—assess the center's performance.

[5] Cooperative agreements are very much like grants in that they are used to transfer something of value to the recipient in order to accomplish a public purpose as authorized by law. The key difference between cooperative agreements and grants is that the federal agency providing the assistance has more involvement with the recipient in carrying out the activity being funded under a cooperative agreement than it does in the case of a grant. Cooperative agreements and grants differ from procurement contracts in that a procurement contract is to be used by an agency when the principal purpose is to acquire property or services for the direct benefit or use of the United States government, as opposed to when the principal purpose is to provide assistance to accomplish a public purpose as authorized by law.

[6] When MEP centers are being established—during the first 4 years of their operation— they receive a larger proportion of federal funds. For more information on the MEP program cost share, see GAO, Factors for Evaluating the Cost Share of Manufacturing Extension Partnership Program to Assist Small and Medium-Sized Manufacturers, GAO-11-437R (Washington, D.C.: Apr. 4, 2011).

[7] In 1998, the program was renamed the Manufacturing Extension Partnership program. In 2004, the program was redesignated the Hollings Manufacturing Extension Partnership in honor of Senator Fritz Hollings.

[8] See 53 Fed. Reg. 27060 (July 18, 1988), 55 FR 38280 (Sept. 17, 1990), 57 FR 1725 (Jan. 15, 1992), 60 FR 29827 (June 6, 1995), 61 FR 20242 (May 6,1996), 65 FR 41634 (July 6, 2000), 66 FR 15219 (Mar. 16, 2001), 68 FR 12890 (Mar. 18, 2003), 75 FR 6355 (Feb. 9, 2010), 77 FR 12563 (Mar. 1, 2012), 77 FR 37653 (June 22, 2012), and 78 FR 21109 (Apr. 9, 2013). Some of these announcements were to establish new centers, and some were to reestablish centers where existing centers had closed.

[9] GAO, Technology Transfer: Federal Efforts to Enhance the Competitiveness of Small Manufacturers, GAO/RCED-92-30 (Washington, D.C.: Nov. 22, 1991).

[10] GAO, Manufacturing Extension Program: Manufactures' Views about Delivery and Impact of Services, GAO/GGD-96-75 (Washington, D.C.: Mar. 14, 1996).

[11] NIST, The Future of the Hollings Manufacturing Extension Partnership (Gaithersburg, Md.: December 2008).

[12] Pub. L. No. 105-309 § 2 (October 1998) (codified at 15 U.S.C. § 278k(c)(5)).

[13] Some centers may still be subject to a lower cost share because they have been in operation for less than 5 years.

[14] The Advanced Manufacturing Jobs and Innovation Accelerator Challenge and the Energy-Efficient Buildings Hub are joint programs with other agencies. The Advanced Manufacturing Jobs and Innovation Accelerator Challenge is a partnership among NIST and the Economic Development Administration within the Department of Commerce, the Department of Energy, the Department of Labor's Employment and Training Administration, the Small Business Administration, and the National Science Foundation. The Energy-Efficient Buildings Hub project is a multiagency effort involving NIST and the Economic Development Administration, the Small Business Administration, and the Departments of Energy, Labor, and Education.

[15] The beneficiary equity standard is commonly used in social science research to design and evaluate funding formulas. See, for example, GAO, Medicaid: Alternative Measures Could Be Used to Allocate Funding More Equitably, GAO-13-434 (Washington, D.C.: May 10, 2013); GAO, Trade Adjustment Assistance: Commerce Program Has Helped Manufacturing and Services Firms, but Measures, Data, and Funding Formula Could Improve, GAO-12-930 (Washington, D.C.: Sept. 13, 2012);GAO, Vocational Rehabilitation Funding Formula: Options for Improving Equity in State Grants and Considerations for Performance Incentives, GAO-09-798 (Washington, D.C.: Sept. 30, 2009); GAO, Maternal and Child Health: Block Grant Funds Should Be Distributed More Equitably, GAO/HRD-92-5 (Washington, D.C.: Apr. 2, 1992); RAND Corporation, Review and Evaluation of the Substance Abuse and Mental Health Services Block Grant Allotment Formula, MR-533-HHS/DPRC, 1997; National Research Council, Statistical Issues in Allocating Funds by Formula, Panel on Formula Allocations, Thomas A. Louis, Thomas B. Jabine, and Marisa A. Gerstein, ed., Committee on National Statistics, (Washington, D.C.: 2003).

[16] NIST could not provide us with documentation to substantiate its estimates of when most centers' received their original cooperative agreement awards. NIST officials stated that NIST's record retention requirements do not require it to maintain files for cooperative agreements awarded in the early to mid-1990s.

[17] Prior to April 2013, the prospective center's budget was evaluated as part of the fourth criterion, management and financial plan, rather than as a separate, equally weighted fifth criterion.

[18] Renewals are contingent upon positive performance evaluations in accordance with 15 C.F.R. § 290.8, taking into account various factors including, as of 2012, whether the center meets the following five qualitative metrics related to the program's strategic plan: (1) the center has an actionable and measurable strategy that is built with purposeful alignment to the MEP program's strategic plan; (2) the center maintains an appropriately balanced portfolio of products and services that address all strategic areas; (3) the center is able and willing to experiment with and adopt new products and services, and has adopted a culture of continuous learning and development; (4) the center works with proactive clients and restarts reactive clients, and creates a sense of urgency that transitions reactive clients to proactive; and (5) the center is willing and able to write success stories that reflect a mix of the strategic plan areas regularly and on-time.

[19] During fiscal years 2009 to 2013, NIST officials told us the following centers' cooperative agreements were opened to recompetition: the MEP center serving Arizona in fiscal years 2009 and 2012, the centers serving the Chicago region of Illinois and central Pennsylvania in fiscal year 2010, the centers serving Rhode Island, Maryland, and Kentucky in fiscal year 2012, and the center serving Nebraska in fiscal year 2013. In five of these eight instances, the existing centers faced financial difficulties that made them unviable and caused them to voluntarily shut down. In the other three instances, the centers decided that their respective organizational missions no longer supported the operation of a MEP center and, as a result, each of these organizations voluntarily ended their respective MEP cooperative agreement.

[20] See GAO-09-798.

INDEX

A

B

C

D

E

F

G

H

I

L

M

W